"An important and overdue biography."   —*The New Republic*

"[An] impressively researched, well-written, and concise account of Thomas's rise from tiny Pin Point, Georgia, to the Supreme Court."
                                            —*Atlanta Journal-Constitution*

"Foskett explores the underlying questions that plagued Thomas regarding the stigma of race and assumptions of unworthiness. . . . The contradictions within Thomas's complex personality are fully on display in this absorbing biography."   —*Booklist*

"Important. . . . Any biographer from here on will need to consult Mr. Foskett's valuable book."   —*New York Sun*

"This in-depth look . . . is a refreshing change. . . . Foskett leaves no stone unturned."   —*Publishers Weekly*

"Engaging."   —*New Jersey Law Journal*

"Foskett guides readers through the shallow stereotypes and explosive controversies to reveal a side of Thomas few would people would ever know. . . . Foskett has given readers keen insight into the burden carried by a black man who chooses to walk a path less traveled."   —*Chicago Sun-Times*

"A warm, intimate portrait of the private man behind the angry mask so often observed on the bench."   —*New York Times Book Review*

"A truly gripping and insightful biography. Written with great narrative verve, Foskett explains the historic rise of Clarence Thomas to the Supreme Court in an unforgettable fashion. Essential reading for anybody interested in both the African American freedom struggle and U.S. legal history."
                      —Douglas Brinkley, bestselling author of *Tour of Duty*

Atlanta Journal-Constitution

About the Author

KEN FOSKETT, an investigative reporter for the *Atlanta Journal-Constitution*, covered legal affairs and state politics before serving as the newspaper's Washington correspondent from 1996 to 2001. Prior to joining the *Journal-Constitution* in 1989, Foskett worked for three years in southern Africa for Save the Children. He is a graduate of Yale as well as the Columbia Graduate School of Journalism. He lives with his wife and son in Georgia.

Perennial

An Imprint of HarperCollins*Publishers*

JUDGING THOMAS

THE LIFE AND TIMES OF
CLARENCE THOMAS

KEN FOSKETT

For my family

First Perennial edition published 2005.

Designed by Jo Anne Metsch

The Library of Congress has catalogued the hardcover edition as follows:

Judging Thomas : the life and times of Clarence Thomas / by Ken Foskett.
    p. cm.
  Includes bibliographical references.
  ISBN 0-06-052721-8 (alk. paper)
  1. Thomas, Clarence, 1948–   2. Judges—United States—Biography.   I. Foskett, Ken.

KF8745.T48J83 2004
347.73'2634—dc22
[B]   2004044935

ISBN 0-06-052722-6 (pbk.)

05 06 07 08 09 ❖/RRD 10 9 8 7 6 5 4 3 2 1

CONTENTS

AUTHOR'S NOTE

My first extended conversation with Justice Thomas took place on Good Friday 2001, outside a small Catholic church two blocks from the Supreme Court. I had been told that he regularly attended services at the church, and I wanted to verify this detail of his life for a tenth-anniversary profile I was preparing for the *Atlanta Journal-Constitution*.

A few dozen worshippers, mostly office workers from Capitol Hill, attended the early morning service. I observed Justice Thomas kneeling in prayer in a pew by himself. After the service I remained behind while he headed for the door. As a Washington correspondent for the *Journal-Constitution* I was used to buttonholing senators and congressmen on Capitol Hill, but I believed morning church service an inappropriate time to corner a Supreme Court justice.

Justice Thomas, however, had lingered outside to talk to several older parishioners, and we practically collided as I rounded the corner. He recognized me immediately from two previous encounters, and instead of turning the other way, as I expected he might, he struck up a conversation. We walked and talked all the way to the Supreme Court and continued our discussion for another twenty minutes in the lobby. I half expected him to invite me up for coffee.

I mark that meeting as the beginning of a relationship that included correspondence, phone conversations, and several hourslong discussions about his life and American history. Although I'd planned to write his biography regardless of his involvement, I believe the book that follows is richer for his having ultimately consented to interviews and spoken candidly about his life.

This book began as a three-part series entitled "The Clarence Thomas You Don't Know," published in the *Journal-Constitution* in July 2001. I am extremely grateful for the guidance and support of former and current *Journal-Constitution* editors Ron Martin, John Walter, Julia Wallace, Susan Stevenson, and Andy Alexander, the Cox Washington bureau chief. Editor Jim Walls remained a source of wisdom and advice throughout the entire project, and I could not have completed the book without his brilliant command of story and narrative.

More than three hundred interview subjects trusted me with their observations and reflections, comprising well over a hundred hours of interview transcripts. I also read several thousand pages of Justice Thomas's speeches, legal writings, and other published material dating to 1974, when he first entered public life as a deputy attorney general in Missouri. I reviewed archival material at the state archives of Georgia and Missouri, the Catholic Diocese of Savannah, the Ralph Mark Gilbert Civil Rights Museum, and university collections of the College of the Holy Cross, Yale University, and Georgia State University. In Washington I examined archival materials at the United States Equal Employment Opportunity Commission, the Alliance for Justice, and the U.S. Senate. I did not have access to Justice Thomas's private papers, letters, memorabilia, or other material that typically might have been available to an authorized biographer.

Because of Justice Thomas's relative youth, many individuals with firsthand knowledge of his life and family history were available to tell their stories. I am particularly grateful for the help of his immediate family—his mother, Leola Williams; his sister, Emma Mae Martin; and his wife, Virginia L. Thomas. All were generous with their time and candid with their insights. Prince Jackson, Sam Williams, W. W. Law, and Ethel Heyward of Savannah were invaluable for their knowledge and memories of Myers Anderson, Justice Thomas's grandfather. I am indebted to several of Mrs. Heyward's children, especially George and Susan, for their memories of Myers Anderson and Justice Thomas. I also wish to thank Dr. Charles Elmore of Savannah State College for his help and his dedication to preserving the history of men and women who built this country.

Dale Couch provided guidance at the Georgia archives, and I am grateful to Carol Van Cleef for sharing with me original source materials relating to her Liberty County relative O. W. Hart. In Liberty County Anna Stevens, Eugene Osborne, and Luretha Stevens invited me into their homes for long discussions about life on the Georgia coast in the early twentieth century. I am grateful to Bill Haynes and Abraham Famble in Pin Point for their childhood memories and knowledge of local history.

Dozens of Justice Thomas's classmates from high school, college, and graduate school consented to interviews. I am particularly grateful to Robert DeShay, Mark Everson, Lester Johnson, Eddie Jenkins, Stan Grayson, Arthur Martin, Clifford Hardwick, Judge Orion Douglass, and Frank Washington. At Yale Law School, former dean Abraham Goldstein and Judges Ralph K. Winter, Jr., and Guido Calabresi graciously opened their doors to me. In Missouri, Clifford Faddis and Richard Wieler went out of their way to accommodate me.

Former Missouri attorney general and senator John Danforth was particularly generous with his time and helped put me in touch with the many fine lawyers and judges who passed through his office in the 1970s. His former chief of staff, Alexander Netchvolodoff, also helped open doors and shared his time and memories.

I wish to thank Senators Orrin Hatch of Utah, Arlen Specter of Pennsylvania, former senator Paul Simon of Illinois, and former congresswoman Patricia Schroeder of Colorado for their time and assistance. Senator Joseph Biden of Delaware declined to be interviewed, as did Anita Hill, but I found material from both in other sources. Senator Biden is quoted extensively by his colleague Senator Specter in *Passion for Truth,* Senator Specter's memoirs. Anita Hill published *Speaking Truth to Power,* an account of her role in Justice Thomas's confirmation, in 1997. For the chronology of the confirmation episode, I relied heavily on the published report of temporary special independent counsel Peter Fleming, Jr., who conducted numerous depositions with most of the principal players. Senator Danforth's 1994 book, *Resurrection,* provides the most exhaustive narrative of Justice Thomas's side of the story, and recollections provided to him were often fresher than what I was able to extract more than a decade later.

I am grateful to many former administration officials in the Bush and Reagan White House, including former attorney general Edwin Meese, deputy William Bradford Reynolds, former White House chief of staff John Sununu, White House counsel C. Boyden Gray, and assistant presidential counsel Mark Paoletta. I also wish to thank Nan Aron of the Alliance for Justice, Wade Henderson of the Leadership Conference on Civil Rights, and Judith Lichtman of the National Partnership for Women and Families.

At the Supreme Court I am grateful to Justices Ruth Bader Ginsburg and Antonin Scalia for their time and insights. Many of Justice Thomas's former clerks were also generous with their time, especially Helgi Walker, John Yoo, and Stephen Smith. Constitutional law scholar Douglas Kmiec gave timely feedback on my understanding of the law and Supreme Court cases.

My wife, Catherine Williams, gave me advice and encouragement that far exceeded customary spousal support, and I could not have completed this project without her. She gave fresh insights to all aspects of my research, offered rich historical context to the narrative, and provided hours of editorial direction. I am also grateful to my father, John D. Foskett, for teaching me to respect people and their political views, and my mother, Margaret Hughes Foskett, for instilling in me the artistic discipline needed for a work of this magnitude. My son, Liam, helped remind me that the most rewarding work is, in the end, fun.

Steve Ecker gave wise counsel, and Mathew Walker and Sally McCarthy were especially kind to open their guest room to me on frequent trips to Washington. Finally, I wish to thank my agent, Heather Schroder, and my editor at William Morrow, Mauro DiPreta.

Although I have included much detail in the book about Justice Thomas's jurisprudence, readers especially interested in the scope and breadth of Justice Thomas's legal writings will want to consult Scott Douglas Gerber's *First Principles: The Jurisprudence of Clarence Thomas.*

I hope, however, readers will share my view that the key to unlocking Justice Thomas's decision making is not dissecting the opinions but understanding the man who wrote them.

JUDGING
THOMAS

———

INTRODUCTION

The United States Supreme Court is home to several hundred support staff, security guards, and maintenance people, who come and go with the daily routine of a large bank. Janitors move around the wide corridors polishing the marble floors, and security guards patrol the hallways to enforce quiet. To employees, including the justices, the building is known simply as "the Court."

The full name is unnecessary. In America the Supreme Court has no equals.

The justices enter through an underground parking deck and work in chambers—somber suites that resemble private libraries. Heavy wooden doors, perhaps two inches thick, insulate the chambers from noise and protect the occupants inside. One of these wooden entries is marked by nothing but a small brass plate at eye level:

JUSTICE THOMAS

A morning in early July 2003 found Clarence Thomas dressed in summer khaki, with a pair of yellow suspenders over a pale blue shirt. Thomas played basketball as a kid, but nowadays he resembles a stocky linebacker more than a wiry point guard, thanks to regular weight lifting in the Court gym. The previous week Thomas had entertained his mother and several childhood friends from Savannah at his home in suburban Virginia. "They always run me ragged," he said in a deep, raspy voice, sounding fatigued.

The Supreme Court's yearly term ended only the week before the family visit, and Thomas looked like he still had not recovered. This term ended on a particularly sour note for Thomas. By a vote of 5 to 4, the Court upheld an affirmative action admissions program at the University of Michigan Law School. Thomas had spent the previous decade building the case against racial-preference programs, and losing the Michigan case was a tough defeat. "I will go to my grave broken-hearted about the dishonesty on race," he said midway through a three-hour discussion about segregation, black power, and his racially charged Supreme Court confirmation.

Pronouncements about Thomas's twelve-year tenure are premature by Court standards; individual records take decades to build. Yet Thomas has already signaled where his legacy lies. The Court's second black justice has waged a sometimes one-man campaign to rewrite the nation's laws on race. Viscerally opposed to classifications based on skin color, Thomas is the Court's leading proponent of what he regards as a color-blind reading of the Constitution and American law. He has voted to take race out of public schools, American universities, the job market, and voting districts.

Thomas thinks racial preferences assume blacks can't compete with whites because of genetic inferiority, what Thomas often refers to as "the presumption." He has been fighting the presumption all his life, even on the Supreme Court. Liberal pundits like to say that the Court's black justice simply obeys Justice Antonin Scalia, as if, Thomas joked, "he was suddenly my master up here."

The preoccupation with race isn't surprising for a man whose life has been indelibly marked by the color of his skin.

Thomas is descended from Georgia slaves who won their freedom during General William Tecumseh Sherman's famous march to the sea. Thomas's grandfather, the man who molded him into the man he is today, was raised by a former slave and later by her adult son. Myers Anderson battled Jim Crow to become a self-made businessman in an era when most black Americans were tied to sharecropping or low-wage jobs.

Thomas's own life spans the end of legal segregation and the beginning of integration. At age sixteen he was the first black student to integrate a Catholic high school seminary. In college he joined the black power movement of the late 1960s, protesting the war in Vietnam and staging campus protests to demand equal treatment for black Americans. He entered Yale Law School as one of the first beneficiaries of an affirmative action program designed to right the wrongs of racial discrimination. Resolved not to let skin color determine his future, Thomas tried to become a corporate lawyer in white law firms but couldn't land a job. John Danforth, Missouri's Republican attorney general, recruited him to promote diversity but also allowed Thomas to do what he wanted, which was practice tax law.

In 1979 Thomas followed his mentor to the United States Senate, but race again changed the course of his life. President Reagan, wanting a black civil rights attorney in his administration, settled on the thirty-two-year-old Thomas, even though Thomas never wanted to be a civil rights lawyer. Reagan then promoted him to chairman of the Equal Employment Opportunity Commission, the federal agency charged with enforcing antidiscrimination laws in the workplace. Thomas was the longest-serving and most controversial chairman in the agency's brief history before President George Bush put him on the fast track to the Supreme Court.

Thomas's confirmation, made famous by Anita Hill's allegations of sexual harassment, was the most racially and sexually charged Supreme Court drama in American history. The 52 to 48 vote remains the closest confirmation victory ever. He is the last conservative named to the current Court and will likely remain its only black member because of mounting pressure to diversify the bench with Americans of Hispanic and Asian descent.

Senior aides to President George W. Bush, whose father put Thomas on the Court, have consulted Thomas about succeeding William H. Rehnquist as the nation's next chief justice. Thomas says he is not interested in the job, and his chances for confirmation in the Senate are problematic, given likely Democratic opposition. And yet the possibility that the court's youngest justice might one day become the nation's first black chief justice remains real. Still in his mid-fifties, Thomas may well serve another quarter century, becoming the longest-serving Supreme Court justice in American history.

Today Thomas is an enigma to most Americans. He is Republican and conservative in an era when most black Americans are Democrats and liberal. He benefited from racial-preference policies but vehemently opposes them. He appears quiet and reticent on a Supreme Court bench dominated by talkers, such as his friend Justice Scalia. The dour, impassive public image contrasts sharply with the man who livens up the room with his explosive guffaws, self-deprecating stories, and bad puns. "He is a good storyteller and has a laugh that is positively infectious," says his liberal colleague Justice Ruth Bader Ginsburg.

Intensely private about his personal life, Thomas cries in public— not often, but suddenly and uncontrollably, revealing a far more sensitive and complex man than the one *The New York Times* once called the "youngest, cruelest justice." He professes indifference to the withering criticism he's received for his conservative views, but he answers in turn with stubborn defiance, insisting that he cannot be bullied.

He has drawn thousands of admirers who respect his conservative philosophy and personal toughness. Yet few justices in American history have attracted as much criticism and condemnation. He has been called a Judas, a traitor to black Americans, a man who has single-handedly undone the victories of the civil rights movement. Today Thomas is an icon for both the right and the left for his beliefs about race, the limits of governmental power, and the protection of constitutional rights.

Clarence Thomas the man has lived a uniquely American life. He was born in a home with no indoor plumbing on a tiny speck of Georgia sand called Pin Point, outside Savannah. His teenage parents divorced before he was three years old. He was shunted between relatives and his mother before winding up with his grandparents. A gifted student and athlete, Thomas studied hard in school and spun an American dream, believing that he might live a fuller, richer life than the one he knew as a poor kid in Georgia. He muscled through discrimination, failure, and disgrace to be where he is today.

Yet if given the chance, Thomas probably would have chosen a different life than that of Supreme Court justice.

"He's dedicated to it, but I'm not sure he likes it," said Harlan Crow, one of his best friends.

Outside the Court, Thomas's life revolves around the grandnephew he adopted in 1997. He leaves the Supreme Court in the early afternoon when Mark is in school so he can be at the bus stop to meet him at three-thirty. It's an odd scene: a silver-haired justice of the Supreme Court surrounded by young, mostly white soccer moms in sports utility vehicles. On weekends Thomas watches Mark play basketball at nearby schools, sitting in the bleachers and cheering with the other parents. Thomas is the only justice with a young child at home, and Mark has become much beloved around the Court, dropping in to see the other justices or playing basketball in the Court gym. Justice Ginsburg called him "the most engaging eleven-year-old I have encountered."

In other ways, too, Thomas isn't a typical Supreme Court justice. He lives for Sunday-afternoon football and NASCAR auto racing. In 1999 he was the grand marshal at the Daytona 500, waving the starter's flag and using his deep voice to command, "Gentlemen, start your engines!"

He drives a limited-edition black Corvette, and in the summer he travels the United States and Canada in a forty-foot luxury touring bus, a kind of houseboat on eight wheels. Wrapped in sheets of steel and painted in a red-and-orange flame motif that sizzles around the side panels, the bus is Thomas's mobile fortress, a suit of armor that allows him the freedom and anonymity he has long craved. Together with Mark and his wife of fifteen years, Virginia, Thomas has driven the bus through more than half the country, stopping nights at RV campsites. At a campground in Michigan, Clarence and Virginia Thomas spent the whole evening around a campfire talking to a pair of postal employees who never asked Thomas what he did for a living and didn't seem to recognize the Supreme Court justice in their midst. "If we can go a day without him being recognized, it feels like a good day," said Virginia Thomas.

Thomas arrives on the job early, often before 6 A.M., an enduring habit learned from his grandfather. He likes the peace and quiet of the Court in the morning, and he does most of his thinking then. His chamber, small by the lavish executive-office standards of today, is a wood-paneled room that looks out over Maryland Avenue. In the wintertime, when the leaves are off the trees, Thomas can just see the U.S. Senate and adjacent office building, where he endured some of the worst days of his 1991 confirmation. Thomas sits with his back to the window in front of an oversize wooden desk, a flat-panel computer screen off to one side. Behind him hangs a large portrait of Frederick Douglass, the escaped slave turned Republican abolitionist whose fiery radicalism Thomas often quotes in his Supreme Court opinions. One of Booker T. Washington, the former Virginia slave who preached self-reliance for freed slaves, hangs over the marble fireplace, which Thomas always keeps lit during the winter. Portraits of both men hung in the segregated library Thomas frequented as a child in Savannah.

Several portraits of President Abraham Lincoln are fixed to other walls, and Ronald and Nancy Reagan smile serenely from a large portrait hanging above a leather-upholstered sofa, facing the portrait of Douglass. "All my heroes," Thomas explained.

Beneath the portrait of President Reagan is a large black-and-white photograph from the mid-1930s depicting Savannah's "Negro" quarter. A dilapidated tenement, near the corner of a dusty, unpaved road, is identical to one a six-year-old Clarence Thomas lived in with his mother in the mid-1950s, when she was working as a maid and he was mostly wandering the streets on his own.

Thomas confided that confirmation to the Supreme Court wasn't the most difficult challenge he'd ever faced. Asked what was, he jerked his hand up and pointed to the photograph. "That was real," he said, looking suddenly grave. "That's what I'm trying to work through."

CHAPTER 1

CLARENCE THOMAS

In the summer of 1955 Clarence Thomas left his mother's tenement in Savannah, clutching a brown paper shopping bag stuffed with everything he owned. The seven-year-old walked with his mother and younger brother, Myers, who carried his own paper bag. The abrupt departure came with no warning and little explanation from his mother. They were going to live with their grandparents, she told them. She would see them when she could, she said. Having seen his grandfather only once in his life, Clarence knew almost nothing about the man who was suddenly to be his guardian. He regarded Myers Anderson as some kind of god, a rich man who lived in a big house and owned both a car and a truck.

He was utterly terrified even before he crossed the threshold.

A tall man, Anderson stood ramrod straight and walked stiffly. Years of chopping firewood, his first business, had chiseled his big hands into callous knobs. One of the few photos of him, taken much later in his life, shows a lean man dressed in a white T-shirt. Anderson's dark black skin, creased and weather-beaten, almost blends in to the mahogany mantelpiece behind him. There is a trace of a scowl across his face. His eyes burn defiantly at the camera, and the muscles in his body look tense, barely suppressing the obvious resistance he felt over being photographed.

Anderson delivered heating oil to black customers in a green Ford truck. Painted letters on the side advertised his business: ANDERSON FUEL OIL CO. The Andersons' two-bedroom house, built just before Clarence and Myers arrived, surrounded the boys with luxuries they had never known: a secure roof over their heads, an indoor toilet, and a bedroom of their own. Anderson bought them clothes and decent shoes. His wife, Christine, the boys' grandmother, made sure they had plenty to eat.

But the comforts came at a price. Under Anderson's roof, the boys lived by his strict rules. He allowed them no break-in period and accepted no excuses. Similarly, nothing prepared Anderson for the sudden responsibility of raising two young boys. He fell back on the only role model he had ever had—the uncle who raised him thirty years earlier in nearby Liberty County—and the rigid discipline and hard work of his youth.

Anderson nicknamed Clarence "Boy," and Myers "Peanut." Boy and Peanut, the two sons he'd never had.

He immediately taught his grandsons to tell time by his clock. Each morning he awoke before the sun rose—"'fo' day,'" he called it—and he worked from "sun to sun." He put his grandsons on the same schedule, rousing them before sunrise every day, even during summer. The morning ritual became so ingrained that Clarence often sensed his grandfather's presence in the predawn darkness before he heard his deep voice. "Get up, Boy," Anderson barked. "Y'all think y'all are rich!" Clarence Thomas cannot remember a single morning of his childhood that he was not up to see daybreak.

Anderson accounted for every minute of his grandsons' time, ensuring that they were busy with chores or homework from breakfast to bedtime. On school days he demanded they be home by three o'clock to help him with his afternoon fuel-oil deliveries. They washed and polished the oil truck; they cleaned his car. They cut the grass and trimmed the hedges. They helped make the cinder blocks that Anderson manufactured in the backyard and sold to neighborhood customers.

On weekends Anderson sometimes drove out to a lumberyard to collect used lumber, which he recycled to build houses. Clarence and his brother pounded out the old nails in the boards and deposited them in a tin bucket; Anderson recycled those, too. Only on Sundays, after church, when he himself rested, did Anderson release his grandsons from his control.

The boys learned quickly that Anderson was never to be challenged in his house. His word was the law. He enforced discipline with a thick belt that he rarely had to use. "He was authority," said Thomas. "You didn't dispute him. And he was very clear about what your responsibilities were, and he meant it. There was no wiggle room. . . . You did what you were told to do.

"When he was talking to an adult, you didn't sit there and gaze at him. You were a child and he was an adult. It was his house and he was the man of the house. . . . He would tell you, 'I rule here.' "

Anderson believed that a man learned by doing, and he deferred nothing to the boys' youth. What he did, they did. The point was driven home forcefully when he roused Clarence and Myers early one morning, just after Christmas 1957, less than three years after they began living with him. Hustling the boys into his '51 Pontiac, Anderson headed south from Savannah toward rural Liberty County, where he had been raised. He typically left the house before dawn on days he traveled, brewing a pot of coffee and pouring it into a Thermos for the drive. Clarence, then nine years old but short for his age, was probably just tall enough to peer through the windows at the blurring countryside.

In Liberty County Anderson turned onto a gravel road and drove for about a mile before veering left down an old dirt track. The car rumbled over the uneven ground and came to a stop in an overgrown field. Anderson's grandfather, Harry Allen, had bought the land in 1893. Now it belonged to Anderson. Anderson called the pastures "the rice fields" after the crop cultivated back "in slavery times."

Anderson stepped out of the car and planted his feet on the sandy soil. The boys watched as their grandfather paced around the open field, thinking. He walked under a tall live-oak tree, then began marking off the ground beneath it: ten paces one way, twenty the other. "We're goin' to build a house," he said finally. He looked at the boys. They knew what that meant. They were going to build the house, too.

Over the next five months Anderson took his two grandsons back to the farmland whenever they weren't in school. He showed them how to pour the foundation and how to mix the sand and cement to make the concrete blocks for the walls. When the blocks were finished, Clarence and his brother, then eight, lugged them one by one to the building site. After the walls were up and the roof was on, Anderson showed the boys how to lay the plumbing for the kitchen, teaching Clarence how to flare the ends of the copper pipes.

Every now and then he looked at his grandson and said, "I'm goin' to teach you to be a man."

They built other houses, too. When the house of a cousin burned to the ground in Liberty County, Anderson paced off the foundation of the new structure while the burnt timbers were still smoldering. On weekends Anderson and his grandsons built screen porches—free of charge—for elderly neighbors. "There was no room for lollygagging, zero," said Thomas. "There was no kind of flexibility at all."

Summertime meant more work, not less, especially after the Liberty house was completed. Anderson moved his family there in the summers and put his grandsons to work reviving the fallow fields. Planting about four acres of crops and vegetables every year, he got the boys up before dawn every morning to help with the plowing, hoeing, and harvesting in temperatures that routinely hovered near one hundred degrees. Under Anderson's direction, the boys strung what seemed like miles of barbed-wire fence around the fields to keep the deer out of the crops. They dug holes and mixed and poured cement into them to set wooden posts, then unrolled the barbed wire and strung it from post to post. "Mendin' fence," that's what Anderson called it. They paused only during lunch and the midday hours when the heat was simply too much, even for Anderson.

When the harvest came in, Anderson put his grandsons to work bagging vegetables for the "po' folk," as he called them. The boys loaded the paper sacks into their grandfather's truck to be dropped later onto the front porches of those struggling to make ends meet. "He'd just give it away," said Beatrice Green, a relative. "He didn't charge nothing."

Sometimes the boys shinnied up a tree to escape Anderson's work demands. Other times they turned the work into games. They raced each other to see who could finish picking a line of beans first, and joked as they went along. But as one neighbor recalled of Anderson, "He put them boys to work."

Anderson often lectured his grandsons on the topic of industry and self-reliance. "You got to do for yourself," he told them. His almost fanatical devotion to work transcended the importance of a secure livelihood. Work gave Anderson a sense of accomplishment and dignity as well as freedom. To be free, a man had to be independent. He had to know how to grow his own food and build his own home. For a black man in the Deep South, self-reliance meant both survival and self-esteem.

Anderson never read well enough to decipher manuals, instruction booklets, or diagrams. Every house he built, every truck engine he repaired, every job he tackled he did without written instructions. He memorized every detail of every task—from how to space the planting of the corn to how to take apart a carburetor. His habit of forcing facts and details into memory stayed with Clarence Thomas his entire life. Thomas's ability to remember names and faces, even those he has not heard or seen for more than twenty years, is uncanny. On the Supreme Court, his ability to recall not just individual cases but specific pages and references inside hundreds of pages of legal briefs astonishes his colleagues. Justice Anthony Kennedy, who sits next to him on the bench, says he frequently relies on Thomas to tell him where he can find citations in attorneys' briefs during oral arguments. Thomas's recall of case details is "photographic," Kennedy said.

Others sometimes told Anderson that he worked Clarence and Myers too hard. But Anderson took advice and orders from no one but his wife, Christine. Tena was the one who told Anderson when he had gone too far, who soothed Clarence and his brother after Anderson's fits of stubbornness or anger. She was, Thomas recalled, "the salve on any wound."

A diminutive woman with delicate features, Tena Anderson kept her husband's books and served him sumptuous meals. Anderson's only indulgence, perhaps, was food, and he enjoyed sitting down to a table piled with his favorites: butter beans, black-eyed peas, and biscuits drizzled in cane syrup. Tena kept out a small bowl of penny candy for the neighborhood children. Anderson adored her.

Summer evenings, Anderson and Tena liked to sit on their screen porch and have a drink before dinner, recalled their neighbor Sandra Cox. "They looked so dignified sitting on that front porch," she said.

Outside the house Anderson loved to joke and kid with his customers or the other youngsters on the street, but at home he felt he needed to project an authority that was uncompromising and severe. Anderson was ill at ease in the role of father, and his family experienced a hardness he showed only to them. Leola Willliams, Thomas's mother, remained frightened of her father her entire life. "He could make me cry just by looking at me," she said.

During the ten years they lived with him, Anderson never once hugged or comforted his grandsons or showed them any affection. "As I got older, I could embrace him as a man," Thomas recalled. "But you couldn't as a kid."

With so little room for leisure, Clarence and his brother never played ball or games with their grandfather. For Anderson, if you weren't working, you were "trifling." He never encouraged playful banter, and he never looked for a show of hands when he decided what everyone should do. "If someone gets mad because I say it doesn't make sense," Anderson was fond of saying, "he has a lifetime to get pleased."

Leola Anderson asked her father once why he was so severe. Anderson blamed his uncle Charles, the man who raised him. "Uncle Charles was raised hard, and he raised Daddy hard," said Leola. "And that hardness just stayed in him."

At home Anderson's intensity could be suffocating, recalled George Heyward, his godson. Years before Clarence and Myers Thomas lived with Anderson, Heyward was Anderson's only "son." Heyward spent hours and hours with Anderson on Anderson's front porch after Sunday mass. Anderson could make him feel like he'd done something wrong, even if he hadn't. After hours of his godfather's sermonizing, Heyward would glance at the clock to determine if it was safe to extricate himself from Anderson's intense gaze. "I think I'll go now," Heyward would say tentatively.

"Oh yeah? Where you going?" Anderson would respond, as if to say, "I know where you're going. You're going to go hang out with those guys on the corner." Heyward would insist he was going home, and Anderson would insist on giving Heyward a ride; that way he could be sure his godson wouldn't be out getting into trouble.

"I don't need a ride," Heyward would say. "I can walk."

Then Anderson would end the discussion with the words that ensured Heyward went straight home. "I know your momma, now," Anderson said. "Don't let me have to call her."

Anderson moralized and lectured constantly, Heyward recalled. "He had very little tolerance for troublemakers and individuals who did not want to work." Anderson knew all the neighborhood trouble spots, and he regularly investigated them, according to Heyward. "He didn't hesitate to come find you."

Anderson had good reason to keep an eye out. Savannah was notorious for a numbers racket that did business on street corners or in "corn houses," places where white liquor ran freely. One neighborhood spot earned the nickname "Bucket of Blood," for the fights that often broke out there.

Yet Anderson's moral vigilance also revealed his beliefs about human nature. For him, the world was composed of good people and evil people; there was a right way to live and a wrong way to live, and no middle ground. He believed that idleness was poison and led to the corruption of the soul. He believed in the literal meaning of the Ten Commandments and told his daughter he would bail her out of jail for anything except stealing. "Because if you steal, you will lie," she recalled Anderson saying.

Perhaps because of his own sexual exploits as a teenager, Anderson became especially vigilant with the young men who fell under his moral leadership. As teenagers, Clarence and his brother were not permitted to attend dances at school, and on weekends Anderson expected them to be home by the time the streetlights came on in the evening. Anderson told Eugene Osborne, a Liberty County relative, that he built the summer house there because he didn't want his grandsons hanging around the city in the summer and getting into trouble, or making what he called "a ruckus."

Anderson also hammered into his grandsons that he expected them to make something of their lives. Failure and giving up were the marks of weak men. Over and over, he asked them the same question: "What are you going to do?" He repeated it so often and with such intensity that it prompted an automatic response inside the boys' heads: "Something."

"His requirement was that you've got to be somebody," observed Prince Jackson, who delivered Anderson's eulogy. "I'm going to allow you in the final analysis to be what you want to be, but you are going to be something."

Apart from school, Clarence found only one escape from his grandfather's relentless authority: Carnegie Library, Savannah's library for blacks. A two-story brick building just three blocks from Anderson's house, Carnegie was the one place Clarence could go without any grief from Anderson. The children's section was in the basement, and as a child Clarence spent many hours there. Quiet and peaceful, the cool basement was a respite from Savannah's heat. Dr. Seuss was an early favorite of Clarence's. Later he discovered the novels featuring Horatio Hornblower, the English sea captain who roamed the Atlantic Ocean. Reading at Carnegie, Thomas first began to imagine a world beyond the city streets of Savannah, a world that even the cynics and the bigots couldn't take away, as he recalled years later. He took books home with him as well, disappearing behind them before bedtime or whenever he could manage to escape. During the summer, Carnegie librarians checked books out for Clarence and gave them to him when he came back into town for church on Sunday. Books sustained him over the long months of summer.

At times George Heyward wondered how Clarence withstood Anderson's constant demands and oversize expectations, believing that he might crack. But Clarence never did. He submitted to Anderson's authority and endured all that was asked of him.

As a young teenager, Clarence began testing the limits of his grandfather's authority, pestering Anderson to allow him more time to play sports, particularly basketball. Anderson's refusal became the nub of increasing tension in the household. Anderson taught Clarence how to drive an old Ford tractor to plow the fields in Liberty. Clarence dutifully performed the work but silently cursed the old man for the domineering control over his life. Why did he always wake him up so early? Why couldn't he sleep in once in a while? Why was he always working while other kids were playing? Why was Anderson so strict?

As an adult, Thomas understood that Anderson's behavior resulted from the pain and abandonment of his own childhood and the tough life he led as a black man in the South. "Here's a man who in spite of all that was productive, successful, and in the end would say, 'I just did my best,'" said Thomas. "I'm not going to apologize for him. I'm not going to make excuses for him, and I'm not going to condemn him.

My brother and I, the only two kids he's ever raised, were able to look each other in the eyes and conclude he was the greatest man we ever knew."

In the end Thomas internalized Anderson's life lessons and personal example in ways that his grandfather couldn't have predicted.

Thomas adopted Anderson's fierce work ethic and relied on that discipline time and again to surmount obstacles that appeared insurmountable. Work gave him the confidence that he could accomplish almost anything. Like the houses they built together, Thomas learned from Anderson to tackle life's challenges one problem at a time. He also copied Anderson's toughness and stubbornness, developing the irascibility that became part of his public persona in Washington.

Anderson's beliefs about self-sufficiency and independence formed the basis for Thomas's beliefs about social welfare programs and the conservative political philosophy that ultimately drew him to the Republican Party. Thomas embraced Anderson's devotion to the "po' folk" and helping those in need with individual acts of charity rather than government intervention.

Although Thomas looks back at his days on the Allen farm with misty nostalgia, the day-to-day reality of the farm was altogether different. Working in that hot sun, rising early every morning, Thomas could think of nothing except getting out. Years later he told a Georgia audience about the hard work his grandfather made him do on the farm. Half serious, half joking, he cracked, "And you wonder why I went to college."

Anderson's relentless domination over Thomas's life created in his grandson a burning desire to succeed. Growing up with Myers Anderson, Thomas yearned to live life on his own terms, free from interference. The yearnings were the first expression of the libertarian political philosophy that defines him today.

Yet Anderson also planted the seeds that make Thomas a deeply conflicted and enigmatic man. Anderson compelled his grandsons to bear slights and hurts without complaint, to force their emotions inward instead of out. As tightly wound as he is, Thomas sometimes can't control his emotions, which tumble out in public unexpectedly.

In their own way, both men were extremists. To this day Thomas has difficulty deciding when enough is enough. He always feels as though he has to be the most committed or the best at whatever he does.

Anderson's inflexibility also made Thomas a stubborn and uncompromising man, a trait that wins him many admirers but often isolates him on the Supreme Court. Like Anderson, Thomas is an absolutist, marking off the world in clearly defined beliefs about right and wrong. The trait often blinds Thomas to nuance.

Thomas's near-photographic memory, inherited and learned from Anderson, is both his blessing and his curse. He remembers every insult, every slight as though it happened yesterday. Deeply antagonistic toward the news media, Thomas can recall the dates of unflattering articles written about him, and the names of the reporters who wrote them. "He don't forget nothing," said his mother. "He do not forget."

Thomas has spent much of his adult life seeking the only thing Anderson never gave him: approval. He stubbornly insists that he does not care what people think about him, but inwardly he craves attention and esteem, say those who know him best.

Thomas clung so strongly to Anderson's beliefs about hard work and self-reliance that they blinded him to other truths about his own path to success. As he moved through life, he could never accept that anything had been given to him. He could never see or acknowledge the breaks that had been offered to him because he was black and had faced a lifetime of discrimination. He developed a sense of entitlement, born not of privilege but purchased fair and square with the sweat of his own brow.

The irony of life with Myers Anderson was that Thomas and his brother were well-off by the standards of Savannah's black community. But Anderson never allowed his grandsons to believe they had anything. He constantly reminded them of the sacrifices he made, that his ancestors had made to get them where they were.

"His goal was to raise us as close to the way he was raised as possible," said Thomas. "He thought that our life was easy compared to his, and he would tell you."

CHAPTER 2

SANDY WILSON

Route 17 crosses the Ogeechee River to the west of Savannah before pointing due south to the Florida line and plunging into some of the richest historical landscape on the eastern seaboard. Guale Indians first roamed and fished the region from their kingdom on nearby Saint Catherine's Island in the sixteenth century. Spanish missionaries established outposts nearly 175 years before the British set foot on Georgia soil. English settlers transformed the lowlands into lucrative plantations of rice and sea-island cotton. The seaport village of Sunbury, now a tiny enclave of waterfront homes, once rivaled Charleston for the wealth that passed through its harbor en route to England.

As boys, Clarence and Myers Thomas rode Route 17 countless times to reach Anderson's farm in Liberty County, eight miles from the old Sunbury village. On their way through the small town of Midway, they passed the white, wooden frame of the Midway Congregational Church, erected in 1792, and the Midway cemetery across the road. White settlers arriving from South Carolina in 1752 built the church as the spiritual bulwark of their new lives in Georgia.

Encircled by a thick brick wall, the cemetery grounds are reached through a heavy iron gate at one end. Giant live oaks, draped in Spanish moss, provide an almost seamless canopy over the grave sites, marked by ornate headstones and sarcophagi.

When he wasn't helping his grandfather on the farm, a teenage Clarence Thomas spent many long hours over the summer battling the forces of nature that daily sought to overrun the old cemetery. Usually working alone, he cut the grass, weeded and trimmed the hedges. It was sweaty, cheerless work. The enormous limbs of the oaks provided some protection from the sun, but the humidity was still stifling. Mosquitoes and other insects buzzed in the shadows, and snakes lurked in the grass. And then there were the graves themselves and the ghosts that a young boy might easily conjure as he worked.

No other place holds as much emotional and sentimental significance for Clarence Thomas as Liberty County. The ties transcend his memories of summer work in the cemetery or helping his grandfather on the farm. Liberty County was in effect the starting point for his unlikely journey to the Supreme Court. Myers Anderson was born and raised in Liberty County and ultimately died there. Thomas's mother was born there before she migrated to Pin Point outside of Savannah. Generations of his family worked and died in Liberty County—first as the property of white men, then as proud Americans scratching out an independent life in a beautiful but unforgiving countryside.

Like many black Americans, Thomas knows little about his ancestry in America. He has sketchy knowledge of a great-great-grandmother named Annie Allen, who was born into slavery before giving birth to a daughter who ultimately became Myers Anderson's mother. Yet Thomas acquired many of his beliefs about hard work, self-reliance, and civic responsibility from these men and women. Under Anderson's iron hand, Thomas learned survival in Liberty County. Like his ancestors before him, he learned how to grow his own food, build a house from the ground up, fix machines that broke, and stick to a job no matter how tedious or backbreaking. The hard work toughened Thomas and made him contemptuous of rewards that weren't earned. Isolated from his friends in Savannah, Thomas spent hours alone in Liberty County, lost in his own thoughts. The boredom, often insufferable, shaped him into the introspective, self-contained man he is today.

Thomas's Liberty County ancestors left him another legacy, one familiar to most southerners, black and white: a lingering, sometimes haunting attachment to the land. When Myers Anderson died in 1983, Thomas inherited the sixty-acre farm, bought by Anderson's grandfather for sixty-five dollars in 1893. The land is a half mile from the pine grove where Anderson and his wife were buried and where Thomas buried his brother, Myers, after he died suddenly in 2000 at the age of fifty.

Every few years Thomas exits Interstate 95 and turns toward the tidewater. The road takes him past fallow fields where his relatives once grew rice, cotton, and corn. He parks on a dirt road in front of the Palmyra Baptist Church, where his brother and grandparents lie buried in the sandy soil, and slips quietly into the cemetery.

Anna Stevens, a Palmyra member, says residents keep a respectful distance as Thomas stands before the headstones of his family deep in reflection, a solitary figure framed by the landscape of his ancestors.

Clarence Thomas's first-known ancestors were a couple named Sandy and Peggy, born near the turn of the eighteenth century. They lived and worked on a plantation owned by Josiah Wilson, a Liberty County planter who died in 1830. The couple had four children, who lived and worked with them and who were also the property of Josiah Wilson. Little Sandy, Sandy and Peggy's eighteen-year-old son, would become Myers Anderson's great-grandfather.[1]

That young Sandy Wilson had survived to adulthood was in itself an accomplishment. In the early nineteenth century Liberty County planters such as Josiah Wilson cultivated mostly rice, the product of gruesome, backbreaking work by slaves such as Sandy. Slaves cut and cleared acres of timber to create the fields, and dug large earthen berms to contain them. They tunneled channels from adjoining rivers and streams to irrigate the fields. They spent hours of the day knee-deep in water and mud. The summer heat, made humid by so much surrounding marsh, was oppressive, and diseases such as malaria and cholera were rampant. Men and women died quick, sometimes painful deaths.

Those who survived developed resistance to disease and acquired endurance for hard, manual labor. One can easily imagine Sandy Wilson as a strong young man with a mind naturally tuned to survival.

Josiah Wilson's estate records give no indication of how he acquired Sandy and his family, or where they may have come from. Josiah Wilson married into wealth, wedding a daughter of Revolutionary War hero Daniel Stewart, perhaps the county's most venerated resident. Sandy and his family may have come with her, or with Josiah from his native South Carolina when he took up residence in Liberty County at the turn of the eighteenth century.

The settlement of Georgia's coastal counties was inextricably tied to slavery. James Oglethorpe, Georgia's founder, banned slaves from the new territory he staked out around Savannah in 1733. But the ban was lifted less than two decades later after white planters complained that they could not compete with planters in neighboring South Carolina, where slavery thrived. With slavery legalized, white planters from South Carolina fanned out along the Georgia coast, grabbing up prime land for rice, cotton, and sugar cultivation. The planters brought their slaves with them by the tens of thousands.

Over time, slaves built enormous wealth for the planters. Whites erected lavish homes for their families and filled them with imported furniture, silverware, clocks, and pianos. Many built second homes away from the coast to escape the hot summer—the "sickly season," as it was called. As legal property, slaves were used to secure loans for their masters, their names embossed on security deeds filed at the courthouse. In a cash crunch, planters hired out their slaves for the season, or sold them outright for hundreds of dollars. The slaves built churches and schools for their masters, while their owners dedicated themselves to a genteel, Christian life.

Sandy Wilson, his parents, and his siblings appear to have remained on Josiah Wilson's plantation while Wilson's executors settled the planter's debts. In February 1833 Wilson and his family were rounded up and taken to the county courthouse in nearby Riceboro to be sold. The auction attracted several of the region's biggest planters, including James S. Bulloch, who at one time owned more than 125 slaves. Bulloch paid $156 to work Sandy Wilson and his family for the next planting season, declining the opportunity to purchase them outright.[2]

Three years later Sandy Wilson's family was torn apart. In a ruthless transaction repeated thousands of times during the era of slavery, they were parceled into lots for the inheritance of Josiah Wilson's five children. Sarah Eliza Wilson, his daughter, drew the lot containing Sandy Wilson and his sister; their parents and other siblings were similarly divided among Josiah Wilson's other offspring. In rare instances, family members separated this way worked close enough that they might still see one another. Often, however, they were scattered so far apart that they lost all contact.

Sandy Wilson remained in Liberty County for several years but then moved to neighboring Bryan County in the early 1840s when his young owner relocated there with her husband, Odingsell Witherspoon Hart, a planter and friend of her late father's.

Wilson, probably one of Hart's most senior slaves, farmed cotton on Hart's five-hundred-acre plantation, Retreat. Slaves generally worked six days a week, from before dawn to after dusk, resting only on the Sabbath.

Sandy Wilson's slave life, however, differed drastically from what other slaves endured in the antebellum South. Slaves in coastal Georgia unofficially held their own property, tended their own fields, and raised their own animals. Although still in servitude, they acquired skills, money, and independence denied slaves in most other regions of the South.

The system evolved from the realities of rice cultivation in Georgia and South Carolina. Planters, many of them absentee landlords from Savannah, learned they could get more work from their slaves by offering special privileges. Certain slaves were permitted to raise and sell animals or cultivate vegetable gardens for their own consumption or sale in town. Some slaves could even own guns for hunting. The awarding of privileges—reserved at first for the slave drivers who managed the field slaves—quickly became advantages that coastal slaves expected as reward for faithful service.

Working by "task," the system was called. The task was literally a unit of land measurement. Five poles made a task, and each pole was twenty-one feet long. Slave owners assigned each of their slaves responsibility for a certain number of tasks. As an extra incentive, the slave owners permitted the slaves to work tasks for themselves. The system differed dramatically from the way slaves were worked in many other regions of the South. Elsewhere, the white overseer determined the day's work and the field slaves fell in line. Under the task system, however, slaves managed their own time.[3]

Detailed descriptions of the task system, in the voices of former Liberty County slaves, survive in complaints submitted to the Southern Claims Commission after the Civil War.

"We all worked by the task, and when that was done worked for ourselves," recalled Samuel Osgood, a freed slave from Liberty County who lived near Sandy Wilson. "I was raising hogs, cattle, and poultry since I was seventeen years old. I used my money to buy clothes and such like that our master would not give, all beautiful nice clothes to wear to meetings."[4]

The system bred a tenacious work ethic. Windsor Stevens said he and other slaves learned early that they could complete two days of their master's task work in one day. "Then the next day was [our] own," he said. Albert Wilson, another former slave, left himself two hours at the end of the day to tend to his own corn, sweet potatoes, and vegetables. William McIver learned how to make tubs and pails from pinewood that he collected from the vast acres of pine groves. He sold his wares, using the money he earned to buy a bacon sow. By selling the piglets, he earned enough to buy a mare and a wagon to hitch to it. "That was the way I got my start," he said.[5]

Liberty County slaves used the task system to slowly extract more land and freedoms from their owners. "I know legally the property was his, but a master who would take property from his slaves would have a hard time," said Joseph Bacon. "Such a master would not get much out of his slave even if he whipped it out of them. And when they had to do that it was poor work."[6]

Slaves with additional skills such as blacksmithing and carpentry could also work for other planters for cash. William Roberts paid his owner eighteen dollars a month for the privilege of hiring out his carpentry services. Anything over eighteen dollars, Roberts kept.[7] "I never failed to make enough to pay him," Roberts said. "I often made a good deal more than enough."[8]

But there were also clear limits. Fearful of competition that could drive down their own profits, plantation owners forbade their slaves from growing cotton, a lucrative cash crop that ultimately supplanted rice as the source of plantation wealth.

Georgia law also forbade the education of slaves. There is some evidence, however, that Sandy Wilson attended church and received religious instruction. O. W. Hart taught the gospel to slaves at a small Baptist church organized in 1835 and likely encouraged his own slaves to attend services.[9] The religious teaching was long on the virtues of obedience and servitude, but it became for many Liberty County slaves one of the few escapes in a weekly life otherwise defined by hard labor.

· · · ·

The 1864 planting season, three years after the outbreak of the Civil War, produced a bumper crop for Georgia's coastal plantations. "I think the Lord gave us a good season because he knew the Yankees were a-coming," Rachel Osgood, a former Liberty County slave, told federal officials.[10]

Since the war's beginning, slaves had tracked its progress through conversations overheard from their owners, or the casualty reports of their sons. Now the plantations around Savannah were alive with news of Sherman's steady march to the sea. Slaves secretly passed information among themselves, usually at night when there was less fear of being overheard. They faced severe punishment if they gave voice to their prayers for Union victory, said Joseph Bacon, who lived on a plantation a few miles from Sandy Wilson's. "The cords were drawn pretty tight," he said. "We had to walk a straight line and do all our work by thinking."[11] Meanwhile, the slave owners spread rumors that Sherman's army would force slaves to pull their wagons, or spirit them off to Cuba to torture and kill them.

With Atlanta in ruins, Union troops were cutting a swath sixty miles wide through Georgia, advancing to Savannah. Sherman's soldiers demolished anything that might aid the rebel cause. They ripped up railroad tracks and wrapped the rails around trees. They burned public buildings. Foraging parties appropriated food stores or destroyed them on the spot. More than inflicting actual damage, Sherman wanted to break the Confederate will, to humiliate Georgia with a show of brute strength and arrogant bravado. Sherman would make Georgia howl, he said at the time. Witnesses to the march invariably seized the same biblical metaphor. *They came like locusts.*

But to Sandy Wilson and tens of thousands of other Georgia slaves, Sherman's advancing host came as a divine instrument of liberation. "The prayer day by day was that they might get through," recalled former slave Henry Stevens, who lived near Sandy Wilson. "The talk was all the time to ask every day, 'How near are they?' " Even in 1873, the year Stevens gave his account to the Southern Claims

Commission, the echo of Sherman's guns still rang in his ears as a harbinger of freedom. "Every time I heard the guns fire, I gave thanks to God and prayed to him that he would let them through and deliver me from this bondage."[12]

In December Sherman reached Savannah, the port city that had for more than one hundred years shipped out the cotton and rice farmed by men like Sandy Wilson, Henry Stevens, and Joseph Bacon. Union cavalry fanned out along the coast, raiding mostly abandoned plantations and taking anything useful. Troops encamped at the Midway Congregational Church in Liberty County just before Christmas. Many of the white plantation owners had already fled to safe havens in southwest Georgia. Instead of rebel sympathizers, the raiders encountered their slaves. The troops plundered corncribs and rice stores and took cattle, hogs, and horses. They also took food, animals, and farm equipment that belonged to the slaves themselves, confiscating personal assets that had taken them years to acquire. "They didn't say anything but 'howdy,' " recalled James Stacy.[13]

But the plunder included a parting gift: You are free now, the soldiers told them.

"When they came and told us we were free, we thought it was too good to be true," said Stevens.

The moment they had dreamed of their entire lives had finally arrived. President Lincoln's Emancipation Proclamation of the year before brought hope to southern slaves,but not their freedom. Sherman delivered on Lincoln's edict.

"I was willing to give all for my freedom," said William Gilmore, another of Sandy Wilson's neighbors. "I would not go back into slavery for twice the amount, or any amount. I value freedom too much to sell it for anything or any price."[14]

Sherman's soldiers spent three weeks camped at the Midway church, methodically hitting every major plantation in Liberty and neighboring Bryan County. "They did not take all in one or two or three days," recalled Joseph Bacon. "They would come and take a little from me and then go to another and take a little from him, and so on till they had stripped the place."[15]

When the soldiers finally reached Retreat plantation, where Sandy Wilson waited for his freedom, they found O. W. Hart, sixty-one years old that December, in bed, sick with fever. The raiders ransacked Hart's plantation house and took the horses, while Hart fled with his daughters to Savannah on foot.[16]

Sandy Wilson likely remained at Retreat, joining thousands of other freedmen who waited out the end of the war in Bryan County.[17]

In Savannah, meanwhile, one of the most historic meetings of the entire conflict took place at a church on January 12, 1865, four months before the war's end. Sherman, accompanied by Secretary of War Edwin M. Stanton, met with twenty former slaves, some of whom had only won their freedom when Sherman took Savannah the month before. The meeting became the basis of Sherman's Special Field Order Number 15, promising freedmen across the South forty acres of land. The order, quickly rescinded, proved a hollow promise.

Yet the testimony of the twenty former slaves gave voice to their powerful yearnings for freedom. "The way we can best take care of ourselves is to have land, and turn it and till it by our own labor," the freedmen told Sherman.[18]

After a lifetime working for other men, Sandy Wilson and his neighbors wanted more than anything else to own and cultivate their own land, to reap the profit of their own work, to never again be dependent on a white man.

In 1867 Hart sold Retreat plantation and moved to Brooks County, Georgia. The same year, Sandy Wilson returned to his native Liberty County to begin life as a free man.

The census of 1870 remains perhaps the most historically significant tally of Americans ever taken. For the first time since the decennial count began, black Americans in the South were enumerated as citizens rather than as property. Suddenly, an invisible nation of people were recorded as flesh and blood, complete with names, ages, and next of kin. For most black Americans, the 1870 census is the first documentary evidence of their ancestry in the United States.

Census takers in 1870 found Sandy Wilson living with his wife, Margaret, who was also known as Peggy, and three children. He'd worked several seasons for white planters, according to tax records.

In 1867 Wilson had exercised his first official act as a freedman by registering to vote. That summer hundreds of thousands of newly freed men across the South registered to vote for the first time under an act of the Reconstruction Congress. Wilson, then about fifty-five years old, and several neighbors walked or rode to the courthouse in Riceboro to declare themselves eligible. Thirty-four years earlier Wilson had stood at the same courthouse as he and his family were hired away to James S. Bulloch for the planting season. But on that day in the summer of 1867 the former slave stood at the courthouse to exercise a right so long denied him.

The following year black freedmen, who outnumbered whites by nearly five to one in Liberty County, again made history by electing two of the first black Georgians to serve in the General Assembly, William A. Golden and Tunis Campbell. Although no one knows for certain, Wilson probably cast his votes for these two lawmakers. Like Frederick Douglass, the great abolitionist whose portrait hangs in Thomas's chamber at the Supreme Court, Golden and Campbell were both Republicans.

In 1871 Wilson opened a bank account. That year his twenty-year-old daughter, Annie, Myers Anderson's grandmother, married a farmer named Harry Allen in a legal ceremony that had been forbidden to them as slaves.

Less than a year later Wilson entered into a contract of historic significance. On March 25, 1872, Sandy Wilson paid one hundred dollars cash for forty acres of Liberty County farmland a stone's throw from where he'd once been a slave. The land had belonged to Oliver

Stevens, the coexecutor of Josiah Wilson's will and the man who purchased Wilson's estate after his death. Now, more than forty years later, Josiah Wilson's former slave had returned to buy the land back.

In the context of the era, Wilson bought more than land on that March day. He bought his independence. He bought security for his wife and family. Sandy Wilson finished out the 1873 planting season in the employment of Monroe McIver, a white planter.

He never worked for a white man again.

Just seven years after gaining his freedom, Sandy Wilson was a landowner.

Although emancipation held great promise for the freed men and women of Liberty County, more than a century of slavery left them a harsh legacy to overcome. In 1870, 3,721 men, women, and school-age children in Liberty County were illiterate, and 92 percent of those were black. Sandy Wilson could not read or write and signed legal documents with an X above his name. Medical care was virtually nonexistent for freed blacks, and disease and illness were a constant threat. Men, women, and children died from pneumonia, typhoid fever, and unexplained "convulsions." Babies died from "teething" and "colic" and "worms," according to the census. Sherman's raiders gave coastal slaves their freedom but also wiped out their assets. Many former slaves ultimately won settlements from the Southern Claims Commission but recovered only a fraction of what they lost. The vast majority of blacks in Liberty County began their new lives with a few tangible assets, such as livestock or horses. Their principal resources were their knowledge of the land and their strength, their endurance, and their ability to produce with their own two hands.[19]

Sandy Wilson farmed his forty acres well enough to feed his family and to pay his taxes. Undoubtedly, he worked as hard as a freedman as he did when he was a slave. But as his neighbor Tony Elliott wrote after the war, at least there was a payoff. "I was glad to get out of my hard labor," Elliott said. "I have to labor just as hard now, but I get the pay for it myself."[20]

Sandy Wilson lived to be an old man, spending more than a quarter of his life as a free man in Liberty County. The year of his death is unknown. But by 1900, tax records referred to "the estate of Sandy Wilson," indicating that he lived into his late seventies or early eighties. Nothing of Wilson's personal life survives, and even his grave site is unknown. Yet the brief outline of his life tells the story of a man determined to claim the rights so long denied him. Like many others, his life was remarkable for the history he lived through. He survived disease and outlived three different owners. He saved enough to buy land. He lived long enough to become a grandfather and then a great-grandfather.

Whatever else, Sandy Wilson was a doer, a man of action at a time when American laws and customs told him he had no rights or freedoms.

Wilson left his farm to his wife, Peggy, and she stayed there until she could maintain it no longer. By 1910, she had moved in with her daughter Annie and her husband. She died sometime before 1920, living long enough to see the birth of a great-grandson named Myers Anderson, the man who taught Clarence Thomas the lessons of his ancestors.

CHAPTER 3

MYERS ANDERSON

Orphaned before he was ten, Myers Anderson never really knew either of his parents. His mother, Laticia Allen, was the daughter of Harry and Annie Allen and the granddaughter of Sandy Wilson.[1] Isaac Anderson, his father, was a Baptist minister from Savannah who occasionally preached services in Liberty County, not far from where Laticia Allen lived with her parents. More than twice Laticia's age, Isaac Anderson impregnated the twenty-five-year-old farmworker while married to his wife of thirty-two years and leading a respectable life in Savannah. In December 1909, five months before Myers was born, he performed the marriage ceremony between Laticia Allen and Peter Williams, a day laborer who lived around the corner from him. At the time, out-of-wedlock pregnancies were taboo and could lead to harsh punishment for the unwed mother. One can only imagine Laticia's predicament when she discovered she was pregnant with the child of a visiting preacher. Perhaps that was why she left home for Savannah. Anderson might very well have helped arrange her marriage to protect both Laticia's reputation and his own.

Isaac Anderson died in late 1917 or 1918, leaving little in the public record. First Macedonia Baptist Church in Savannah appears to have closed or splintered with his death. Laticia died around the same

time, shortly after giving birth to a daughter named Lizzie, a half-
sister to Myers, who was also raised in Liberty County.

Myers Anderson's confusion about his birth extended well into
adulthood. On different records, he spelled Laticia's name two different ways and listed his birth year as either 1907 or 1910.[2] Doubtless,
he learned some details about his parents as he grew older, but he never
talked about them to anyone, perhaps out of a sense of shame. The
only real parents he ever knew were his grandmother Annie and his
Uncle Charles, the man who played the greatest role in shaping him.

At the turn of the twentieth century the extended Allen family lived
together on Harry Allen's farm in Liberty County, farming and fishing. Harry and Annie were both born slaves; she in 1850 or 1851 and
he in the late 1840s. Allen farmed alongside his father-in-law, Sandy
Wilson, until 1893, when he purchased sixty acres of timberland less
than a mile from Sandy Wilson's forty-acre spread. Several hundred
black families, all descendants of coastal slave owners, owned farms
there, living separate, independent lives. The land was considered
largely useless because it didn't back up to a river, creek, or the marsh,
conditions ideal for tidewater rice farming. But the soil was good, and
Allen and his neighbors farmed sweet potatoes, cotton, butter beans,
corn, peas, and sugarcane, even some rice. The Sunbury Baptist
Church was the anchor of the community, along with a small, one-
room schoolhouse. Harry and Annie had at least eleven children, including Laticia, who was also called Ticia.

Harry Allen died in 1911, leaving Annie the aging matriarch of
the family. Census takers found Myers living with her in 1920, listing
his age as nine. Doubtless, the arrangement was as much for Annie's
benefit as for his; younger children were frequently detailed to live
with older grandparents to help them with chores. Annie died on October 18, 1924, at the age of seventy-four. She was buried in a simple
grave next to the Sunbury Baptist Church. The eighty-year-old grave

site is perhaps the oldest surviving burial ground in Clarence Thomas's family.

After Annie's death, Myers moved in with his Uncle Charles. At only fourteen, he had already been touched by three deaths: those of both his parents and his grandmother.

Charles and his brother Jim Allen, two of Harry and Annie's oldest children, inherited the family farm after Annie's death. They led different but complementary lives. Jim, a farmer, took after his father. Charles was a "river man." The salt marshes surrounding the irregular coast of Liberty County were rich in seafood—oysters, shrimp, blue crab, and fish. Uncle Charles knew the marsh well and handled a boat with ease. In his years of fishing, Charles developed an almost supernatural ability to predict weather patterns, forecasting storms hours before they materialized.

As a boy, Myers alternated between the two worlds of his uncles. He farmed with Uncle Jim and fished with Uncle Charles. Both demanded long hours of work. Charles rarely saw his children because he rose before dawn and returned after dark, having hauled his catch to neighboring towns for sale. The farming life was not much different. James was in his fields at the first hint of light, Myers and the other children in tow. He paused only during the hottest part of the day. When he wasn't working the river, Charles farmed alongside his brother.

Sometimes Charles hired himself out to the white ship captains who transported goods and people up and down the Georgia coast, according to Rosa Allen, his youngest daughter. But not Uncle Jim, Rosa recalled. He would work for no man, she said, except himself.

The two men carved out a life that was almost entirely self-supporting. They kept cows for milk. They made their own butter, crushed their own sugarcane for syrup, and kept a smokehouse for meats. They jarred vegetables and produce for the winter. They hunted deer and opossum. They knew both animals and machinery and could fix nearly anything that broke. They built their own houses and dug their own wells. They wanted to be independent. And they were.

But their independence came at a steep price for their children. Self-sufficiency depended on available hands to work in the fields and help with the chores. "Childhood" was a brief interlude that lasted no longer than age six or seven. By that time children were detailed to the fields. The youngest became living scarecrows, chasing birds away from the ripening crop. Older children were given hoes to keep the weeds at bay, and boys learned how to steer mule-driven plows. Parents allowed children out to play only on Saturday afternoons after they had cut the firewood used for cooking and heating. Sunday was spent almost entirely at church.

Parents tolerated no misbehaving of any kind, a point driven home to Anderson one day on Uncle Charles's boat. In a fit of temper Charles whacked him with an oar and knocked him overboard, punishment for horsing around on the water. The harshness conveyed a lesson about Georgia life in the 1920s: For a black man, the line between life or death was as flimsy as a rowboat in rough seas.

Years later Myers Anderson told his friend Sam Williams that he really kept only one memory of his childhood: hard work.

School was a luxury that many families could not afford, particularly for the boys. The one-room schoolhouse Myers Anderson attended for two or three years first opened after the Civil War as a collaborative effort between local Baptist ministers and the African Missionary Society, a northern religious organization. The school went only as far as the seventh grade, but completion depended enormously on a stable home life. Luretha Stevens, a contemporary of Anderson's born in 1920, recalls her mother's premature death as the worst event of her brief childhood; from that point on, Stevens said, the household chores fell to her. For a time Stevens and her brother shared the responsibility. She went to school one week, and he the next. But then she dropped out. Her father, she said, didn't believe an education held any relevance for farmwork. What mattered was what you could do with your hands.

But if the engine that drove the community of Myers Anderson's youth was hard work, the oil that kept it running smoothly was the commitment of neighbors to help one another. Anderson's surviving contemporaries cherish their memories of how neighbors routinely dropped by one another's houses with a "mess of greens" or a "mess of beans" or "a mess of fish," recalled Luretha Stevens. Every parent in the community had the authority to discipline any child. When a man was sick, he could count on his neighbors to help tend his field. If he died, he knew that someone would take in his children. Crime was almost nonexistent. That notion of civic responsibility remained with Myers Anderson his entire life.

"He came from the old school," said Anna Stevens, who grew up on a farm near Anderson's. "He'd say, 'Come on, the peas is ready to be picked, come and get them.' Whatever he had he would share."

As a teenager, Anderson worked mostly with his Uncle Charles on the water. Allen gave the money for the day's catch to his wife to keep in a small chest in the kitchen. But he kept a record of Anderson's share, and he told Anderson that when he felt he was ready to strike out on his own, the money he earned was his. "All you need to do is tell me," he told Anderson.

Anderson stopped working with his uncle in his late teens and took a job with a white contractor who had a house-moving business in Savannah. Early every morning the man picked up Anderson and one or two other men and drove them thirty-five miles to Savannah for a day's work. When the contractor moved to Savannah, Anderson faced a decision: Stay in Liberty County and make a life for himself on the land or the river, or move to Savannah. He chose the latter. Charles Allen went to the small chest in the kitchen and gave Anderson his savings. "He make a man out of Myers before he leave here, yes he did," said Rosa Allen. "He make a man out of Myers."

Anderson left Liberty County at the age of twenty, determined to cut his own road in life. He had little money and virtually no formal education. His two greatest assets, perhaps, were ambition and a seemingly limitless tolerance for hard work. He needed both to survive what lay ahead.

Savannah in 1930, the year Anderson likely arrived, boasted a dynamic mix of races and ethnicities. German and Irish immigrants and black islanders from the Caribbean coexisted with the moneyed white aristocracy. Many of the city's black residents were descendants of the freedmen who followed Sherman to Savannah on his march to the sea. Although there were solidly segregated neighborhoods, there were also many areas where whites and blacks lived in close proximity. Ben Lewis recalls that the street of his youth in the 1920s was black on one side and white on the other. "I played with white kids as much as I played with black kids," he said. John White, born in 1924, grew up on a street where every other family was white or black. Duplexes often had a white family below and a black above, White recalled.

The city was also home to an established class of black tradesmen, educators, ministers, and a handful of doctors and dentists. The *Savannah Tribune,* founded after the Civil War, published the news of the city's black community. Georgia State Industrial College for Colored Youth, the state's first black college, attracted scores of talented educators. The cotton exchanges and tobacco warehouses, along with the bustling shipping industry, provided jobs to hundreds, perhaps thousands of unskilled black workers. The city's black elite lived in old Victorian homes near West Broad Street, the heart of the black commercial district. The street was anchored by the Wage Earners Loan & Investment Company, the first black-owned bank in the nation to exceed one million dollars in deposits. The concentration of black banks, insurance companies, and trade shops earned West Broad the name of "Black Wall Street." Vaudeville troupes featuring black entertainers played to packed houses at black-owned theaters. Socially conscious blacks formed exclusive clubs to hold balls and showcase their debutantes. Scores of black societies, lodges, and fraternal organizations thrived in Savannah. At funerals members dressed in uniforms and feather-plumed hats and marched beside horse-driven caissons to Laurel Grove Cemetery.

Anderson arrived in Savannah, however, at the beginning of the Great Depression. The Wage Earners bank failed in the collapse, wiping out hundreds of thousands of dollars in savings accounts. More pernicious was the sudden replacement of black longshoremen and ship hands by unemployed whites. W. W. Law, a friend of Anderson's, said his father lost his mechanic's job in 1932 and was replaced by a white man. He died shortly thereafter. "He lost the will to live," said Law.

Ethel Heyward and her husband, who knew Anderson well, returned to Savannah in 1930 to find the city dramatically changed. "We had no food and no jobs," recalled Heyward, who was born in 1912. "If your neighbor had food, they would call you. If they had bread they would call and share it with you. That was the way we survived." Law recalled days as a boy in the 1930s when he stood with his mother in bread lines, rising before dawn and waiting until mid-afternoon in the hot sun. He bitterly recalled instances when those appointed to distribute the food—black church members no less—supplied their friends from the back of the church and left those waiting out front with nothing.

Savannah's diverse mix of people spared the city from some of the violence toward black residents that gripped other parts of the South. Yet white supremacy and intimidation were still daily facts of life. John White, Savannah's first black police officer, recalled that the corner grocery store of his youth in the 1930s featured the severed hands and feet of a black man displayed in the glass case under the counter. The store owner's message was clear, said White: "That could be you, boy."

The Ku Klux Klan marched frequently in Savannah, and Klan members dominated the police ranks of the 1930s, '40s, and '50s, according to White, who joined the police force in 1947. As a boy White recalled seeing a white police officer crush his nightstick into the skull of a black man simply because he grabbed his arm to report a fight. The force of the blow knocked the man's eye out. "All because he didn't want a black man to touch him," said White bitterly.

For several years Anderson continued to work for the house mover, living in a boardinghouse and saving his earnings.

Then in the 1930s he fulfilled his dream: He went into business for himself. In that era, city dwellers depended on two things to keep their homes comfortable: coal or wood for heating, and blocks of ice for refrigeration. Using his savings, Anderson bought an old truck to start a coal, wood, and ice delivery business. Still in his twenties, he could not read or figure numbers, two deficiencies that made his business dealings difficult. But Anderson knew how to work. He began his ice deliveries before dawn, depositing twenty-five- to fifty-pound blocks in the storage chests that people kept outside on their porches. At night he drove to Liberty County to purchase firewood from a wholesaler, returning to Savannah well after midnight.

The delivery business suited Anderson well. Although he continued to farm in the summer down in Liberty County, he confided to friends that he never wanted to be a farmer. He was, by nature, a kinetic man, unable to sit still. The constant movement of the delivery business satisfied his need for motion. Behind the wheel of his truck, Anderson was in control. He set his own hours and followed his own orders. "He wanted to be independent," recalled W. W. Law, twelve years Anderson's junior. "And he was."

The first real payoff for Anderson's hard work came on November 21, 1938, when he bought a vacant lot on East Thirty-second Street for $450. Then he built himself a house. The two-bedroom structure was modest, but at the peak of the Great Depression it was a monumental achievement.

The year before, Anderson's Uncle James had sold Sandy Wilson's farm back to the family of the white planter who had owned it before the Civil War. Anderson saw the transaction as a humiliating capitulation: the descendants of slaves selling their hard-earned land back to the family of their slave masters.

By the 1940s, when homeowners began replacing their wood- and coal-burning stoves with oil-burning furnaces, Anderson switched to delivering fuel oil. For a while he delivered both fuel oil to his fuel oil customers and coal to his coal customers, all the while keeping up with his ice deliveries.

Anderson's business competed not only with other black delivery drivers, but with white ones as well. Most skilled jobs in Savannah were simply not open to blacks. At the rail yard blacks were the brakemen but never the engineers. On the river they were the longshoremen but rarely the riverboat pilots. Anderson's business, however, was open to both white and black. Anderson was forever teaching his grandchildren that they were no different from whites, and they could compete with whites. For Anderson, this was more than idealistic rhetoric. He had competed with them successfully.

Yet Anderson knew that the playing field wasn't level. Years later he confided to Prince Jackson that he always knew he was paying a few cents more for fuel oil than his white counterparts, who could charge a few pennies less to their customers. But Anderson offered his black customers something that a white dealer rarely did: credit. He filled the oil tanks of his customers even when they couldn't pay him. It was the act of kindness for which he is most often remembered. Sam Williams, a close friend, remembers seeing hundreds and hundreds of dollars in uncollected deliveries in Anderson's account books. Some customers owed as much as three hundred dollars. "That was a lot of money back then," Williams said. Once he was old enough to ride the fuel-oil truck with his grandfather, Thomas became aware of many such customers. Thomas remembers his grandfather once leaving a house with tears in his eyes. Just fill it and let's go, he told his grandson. There were two things that Anderson could not bear to see: people going hungry and people going cold. As a boy, he had known both.

Anderson also attempted to sell the cinderblocks he made in his backyard to the city, but his friends say city officials claimed they weren't good enough. Sam Williams believes a white-owned company elbowed Anderson out and then obtained the lucrative city contract for itself. Anderson continued, however, to do a good business selling his blocks to black customers and used his earnings to buy lots and build houses that he then converted into rental properties. He acquired nearly a dozen lots in the 1940s. The properties ultimately played a pivotal role in the civil rights movement that was yet to come.

Anderson's silent partner in all his business dealings was his wife. He married Christine Hargrove in 1933, three years after arriving in Savannah. Christine, or Tena, as she was affectionately called, came from Richmond Hill in neighboring Bryan County. She'd been an unwed mother before she met Anderson, but her child lived only two months, according to Anderson's daughter, Leola. Unlike Anderson, Tena attended school until the sixth grade and could read and write. Anderson depended on her to keep track of his earnings and maintain his books. "She was right there beside him," recalled Anna Stevens, who knew the couple when they retired to Liberty County. "She handled the money. He made it. But he put it in her hand to take care of." Indeed, Tena never held a job outside the home, a luxury that only the most self-sufficient black families could afford.

Despite his relative wealth, however, Anderson insisted on living modestly. A cluster of Jewish shop owners ran small businesses on East Broad Street near Anderson's home. Anderson studied their discipline and thrift, frequently extolling Jewish industry as a model for black tradesmen like himself. His devotion to industry bordered on obsession; too many comforts made a man sluggish, he thought. The trucks he drove came installed with heaters; Anderson took them out. A heater was a luxury that made a working man lazy, he said.

In the early 1940s Anderson met a man who was to have a profound influence on him and ultimately his grandsons. Sam Williams, who credits Anderson with encouraging him to get into the delivery business, introduced Anderson to the Catholic church and urged him to become active in the Savannah chapter of the NAACP. The son of immigrants from the Caribbean, Williams had attended segregated schools in Savannah and could read and write. The two saw each other at their refueling stops at the oil depot and shared business tips. Williams says that when he met Anderson, he saw that he had taught himself to recognize words on delivery receipts and customer addresses but had never learned how to figure numbers. Williams taught him a system of rounding numbers to make addition and subtraction easier. Anderson absorbed it all. As he taught Anderson math, Williams said, he began to appreciate the sharpness of his intellect, his ability to absorb a wide range of concepts and quickly commit them to memory.

In addition to mastering carpentry, plumbing, and farming, Anderson was a crack mechanic. He could take apart a truck engine and put it back together flawlessly—all without the aid of manuals. Williams recalled that when his own truck broke down, Anderson offered to fix it. He removed the engine and dismantled it. He memorized the location and function of each part before putting the engine back together. "That truck lasted for five years after that," Williams said. "I never seen anything like it."

In 1949 Anderson confided to Williams that he was unhappy with the local Baptist church that he had been attending. Williams suggested Anderson join him at Saint Benedict the Moor on East Broad Street, about eight blocks from where Anderson lived. Saint Benedict's was the first of three Catholic parishes serving Savannah's tiny population of perhaps one thousand black Catholics.

Dr. Prince Jackson, who delivered Anderson's eulogy, believes Anderson was probably deeply moved by the church's presence in the black community. Although the parishes were black, the priests at Saint Benedict's and the other churches were all white. White nuns taught at the church's three schools for black children. Many black Savannahians admired the white priests for their selfless dedication to the black community, Jackson said. "Here was a white guy saying, No, you are a creature of God," Jackson said. That message matched Anderson's own view of God's creation of whites and blacks as equals.

Anderson may have been drawn to the Catholic church for another reason. At the time, potential members were required to attend several months of catechism classes. For a man who had never really attended school, this opportunity was probably powerfully appealing.

Anderson was baptized at Saint Benedict's on May 8, 1949, and threw himself into church life. He attended mass every Sunday and was often at the church for meetings during the week. Even when he retired to Liberty County he returned every Sunday for mass. He befriended the nuns who lived at the convent next to the church, and he filled the church's tanks with fuel oil. He routinely dropped by the convent with a basket full of fresh vegetables from the farm. On weekends he repaired the church building. He rebuilt its front steps and installed a hand railing, impressing church members not just with his handiwork but also with his religious devotion.

"He was one of the first people who told me that even if you are alone, you are in the majority if you've got God on your side," recalled Jackson. "With God on your side, nothing was really impossible. All you had to do was continue to work."

In the men's Bible-study group, Jackson recalled, Anderson returned to one topic over and over again: How can Christians apply the teachings of the Bible to everyday life? The question helps explain Anderson's extraordinary acts of charity and kindness. Williams recalled that Anderson would never speak ill of someone, even if he had cause. When Williams fumed against the latest outrages of the white establishment in Savannah, Anderson was always the first to ask, "And what of the black man? Is he so perfect?"

Anderson took a keen interest in the neighborhood children. He made sure they had pencils and paper in school. He kept treats and pockets full of loose change, and children ran out to greet him on his delivery routes. "Mr. Anderson! Mr. Anderson!" they shouted.

And Anderson reached into his pocket and gave out a nickel or maybe a dime.

In Liberty County, where he returned frequently year-round, he organized building brigades to construct homes for the community's elderly. He gave money to the Allen family and was always there to help. "He never forgot us," said Rosa, Charles Allen's daughter. "We didn't have to worry about nothing."

The Bible preached that it was better to give than to receive, and Anderson not only believed it but sought to live it. Over his lifetime he read the entire Bible twice, cover to cover, an astonishing achievement for a man who struggled with reading. As a boy, Thomas watched his grandfather sitting with his worn copy of the Bible, laboring over every word. Anderson read the passages aloud, haltingly, and then committed his favorites to memory. He could pull entire verses out of his head and recite them in his Bible classes, or when he wanted to make a particular point. "The Lord is my Shepherd" was one favorite.

Anderson also studied and took the teachings of the Catholic church to heart. Throughout his life, he fervently opposed abortion and believed it was a sin.

And yet as a man Anderson was full of unexplained contradictions and mysteries. No one knew those better than his only surviving daughter, Leola.

Clarence Thomas's mother, Leola, guards a sixty-three-year-old memory. She remembers being bustled off to a church in Liberty County to attend a funeral when she was twelve. "I didn't even really know what a funeral was," she recalled. The small church was crowded with people. Some were crying. A girl Leola had never seen before lay in an open coffin at the altar. The girl, only fourteen, had died giving birth to a baby. A tall, older man whom Leola did not recognize stood at the head of the casket.

The dead girl was her sister, she was told. The man, Myers Anderson, was the girl's father. He was also Leola's father. It was the first time she remembers seeing him.

Myers Anderson was just seventeen when Leola, his second daughter, was born in 1928. His first, Mildred, had been born in 1926 and conceived when Myers was just fifteen and living with his Uncle Charles in Liberty County. Myers didn't marry either girl's mother, and when he left Liberty County as a young man, he left them both behind. Mildred, who died two days before her fifteenth birthday, was the sister Leola never knew.

Leola's mother, Emma Jackson, worked as a laborer on a Liberty County plantation. Leola remained in the care of Emma's mother while Emma worked in the fields. In the early 1930s Emma moved with her young daughter to Savannah. She married in 1934 but disappeared a year or two later. There is no record in Georgia of her death, and Leola has no idea where her mother is buried.

Leola was raised by Emma's sister, Annie Jones, who lived in Pin Point, just outside Savannah. Myers Anderson lived just eight miles away, but Leola did not even know he was alive. At her sister's funeral, she asked her father why she wasn't living with him. "When you get old enough, I'll tell you," he replied. Leola said she counted the days until she could learn the answer.

As a young teenager, Leola took the bus to Savannah and found her father chopping firewood behind his house. She asked him again: Why? Anderson told his daughter that he wanted to take her in, but her aunt Annie refused to let her go. The encounter began a stormy relationship between father and daughter that lasted until Anderson's death.

Even now Anderson's indifference remains a source of bitterness for his only surviving daughter. "In Savannah, all you hear is, 'Oh, Mr. Anderson, oh, he was so great,'" said his daughter years after his death. "What did he ever do for me? Nothing."

• • •

The coastal environs of Savannah are a labyrinth of rivers, creeks, and streams that feed into the Savannah and the Little Ogeechee Rivers. The waters swirl lazily around fingers of mainland and dozens of islands, large and small, creating an enormous salt marsh rich in seafood and wildlife. At the fingertip of the mainland, hidden behind a tall stand of pine trees, lies the tiny community of Pin Point.

Freed slaves from the South Carolina and Georgia coast settled Pin Point in the 1880s. They fished the rivers or worked in the large plantation estates up the Moon River, living in small clapboard houses clustered along a vast marsh. Outsiders often called Pin Point residents Geechees, after the nearby Little Ogeechee River. Up the coast in South Carolina, the islanders were called Gullah because of their unique language. Later, Gullah/Geechee came to describe the distinct culture of the Georgia and South Carolina coasts. Many Gullah/Geechee traditions originated in Africa and were still in practice when government anthropologists stopped in Pin Point in the late 1930s. One woman described how a few of the older residents practiced "conjuhin," or conjuring, to guard against evil spirits. Interviewers learned of a Pin Point fisherman recently killed by an alligator and were told of a recipe for "mojoe," a potion to ward off enemies.[3]

A dirt road ran the length of the community. Residents drew their water from a communal well and used outhouses, living conditions that remained unchanged until the 1950s. As in Liberty County, the folks of Pin Point relied on themselves and one another to survive.

"Most of the homes didn't have locks," recalled Bill Haynes, a contemporary of Leola's who was born in Pin Point in 1926. "You didn't have to worry about anybody taking anything. We were poor, but I guess we didn't know it."

Although the Great Depression fell hard on Pin Point, in some ways the community was slightly better off than others in the South.

People grew their own vegetables—okra, collard greens, tomatoes, and corn. The men trapped raccoons and opossums. The river was full of croaker, whiting, bass, and flounder, not to mention crabs and oysters. A nearby seafood cannery provided meager income. "They worked all week and had very little money to show for it," recalled Haynes.

Leola's childhood in Pin Point was typical. She walked to and from Haven Homes Elementary, the county's school for black children. When she was nine, she was already picking crabs and shucking oysters at the Pin Point cannery, earning five cents a pound. The shells were sharp and jagged, and they cut her young girl's hands. She attended the local Baptist church, which she still belongs to.

But Leola's childhood memories are colored by the harsh treatment she received from her aunt. Annie Jones kept house for a wealthy white family in a mansion upstream from Pin Point. Each day she assigned Leola chores to do around the house. Sometimes Leola played rather than worked, and Annie whipped her for shirking her chores. To escape the whippings, Leola climbed the palmetto trees when she saw Annie running for the switch. Other times, she sat on the back porch, hugging herself and singing a made-up song. "A motherless child sees a hard time when Mother's gone," she sang over and over. Sometimes Annie stopped the whippings when she heard the song, Leola said.

Annie Jones's husband, Ed Jones, was the one bright memory in Leola's childhood. He was a kind man, Leola said, and often intervened when his wife was hard on her niece. When Leola was a teenager Uncle Ed devised a signaling system so she could escape the house at night. Annie forbade her niece from going out after dark or attending community dances. On dance nights Leola climbed into bed fully clothed and waited for Uncle Ed to cough loudly, the signal that her aunt was asleep. Leola crept out of bed and climbed out her window into the hot summer evening.

On one such night out Leola met M. C. Thomas, a teenage farmhand at the Bethesda Home for Boys, just up the river from Pin Point. The boys' home was one of the nation's first orphanages, founded in 1740. In the 1940s it was still a home for white orphan boys. M. C. Thomas helped his father cultivate the crops and tend the animals that fed school residents.

M. C. Thomas's journey to Pin Point began in the unforgiving cotton fields of Johnson County some 120 miles to the northwest. He was the son of November and Minnie Thomas, who began their own lives under the grinding cycle of sharecropping that entrapped hundreds of thousands of black southerners. At the turn of the century November's father lived with his nine children on a white man's farm outside of Wrightsville, Georgia, in small shacks provided by the owner. They grew cotton, corn, and sweet potatoes, but the farmer charged so much for rent and livestock feed that they rarely earned a profit.

Willie Mae Thomas, November's sister-in-law, said the Thomas family moved every year to a different farm, bouncing between Johnson and neighboring Laurens County. Changing farms gave the Thomas family the illusion of independence, but in reality life never changed. Willie Mae said it was impossible for anyone to get ahead because the landowners took most of the crop. "We didn't get much off it," she said. "We paid our bills and did what we could." The worst, Willie Mae recalled, was the "Hoover days" of the Depression, when families often did not have enough to eat. "Some people walked the roads like dogs, didn't have nowhere to stay, didn't have nothing to eat," she said. "It was rough. It was rough back in them days. I mean, bad."

November Thomas was determined to break the cycle of sharecropper dependence. In the late 1930s he packed up his family in the dead of night and moved to Savannah to take a job at the boys' home. Within weeks, however, his former landlord found him and accused him of running off with unpaid debts. To settle accounts, the farmer loaded Thomas and his family into his truck and brought them back to Johnson County. Thomas left for Savannah after the next season, but the white farmer once again claimed that Thomas owed him a crop. "Then they left the third time, and the man didn't come back no more," recalled Alice Thomas, M.C.'s sister-in-law.

November Thomas and his family lived on the grounds of the Bethesda boys' home. While the older boys joined the armed forces and went off to fight World War II, M.C. stayed behind and helped his father work the school's farm. Just after the war, he met Leola.

Leola Anderson became pregnant with her first child when she was sixteen and M.C. Thomas was just fifteen. Her daughter, born in 1946, was named Emma, after the mother Leola barely knew. Myers Anderson was furious when he learned that his daughter was pregnant and ordered her to marry the baby's father, even though Leola told him she was too young. "If you make your bed hard, you are going to lay in it," he told her, but Anderson's hypocrisy galled his daughter. "He expected me to be the perfect child, but he didn't have to be," she says.

Leola had few options. The Baptist church in Pin Point expelled her when they learned she was pregnant, and she dropped out of school after completing the tenth grade. She married M.C. Thomas, then seventeen, in January 1947.

Eighteen months later, on June 23, 1948, she gave birth to a baby boy. The child was delivered by a midwife inside Leola's tiny house. She named the boy Clarence for reasons that she can't recall today.

Leola went back to work at the cannery while she was pregnant with her third child, now earning fifteen instead of five cents a pound for picking crab. Another son, Myers, was born sixteen months after Clarence. Leola was the mother of three children, the oldest just a toddler, before she was twenty-one years old.

M.C. Thomas earned twenty-five dollars a week at the boys' home, and with three children to clothe and feed, money was extremely tight. Myers's birth in 1949 strained the family to the breaking point, and Leola seemed unprepared for the constant attention that three small children required.

Three years after their marriage, M.C. sued for divorce, claiming "acts of cruel treatment." Thomas claimed his young wife neglected their children and home, that she "would absentee herself away from their home at all hours of the day and night until the early hours of the morning in the company of people with whom she should not be."[4]

At the time Clarence Thomas was only two and a half years old. Leola never responded to the complaint, and in March 1951 a Chatham County judge granted the divorce. He awarded custody of their three children to Annie Jones, Leola's aunt, and required M. C. Thomas to pay ten dollars a week toward their care and feeding.

Years later, in her early seventies, Thomas's mother said she left M.C. because he was running around with another woman who became pregnant. "You ruined my life," Leola said she told her ex-husband. "I told him that was the end. I [had] three babies to raise, no husband, no income, had to go to court, and then he fought me in the court to take care of them."

Annie Jones took custody of her niece's children in 1951. Leola moved to Savannah and became a maid, returning to Pin Point on the weekends to see her children. M. C. Thomas remained in Savannah but almost never visited his offspring. He saw Clarence Thomas only once again before he moved north to Philadelphia in the late 1950s, after Thomas and his brother left Pin Point. He never explained the circumstances of the divorce to his children, Thomas said.

Thomas's mother says neither of her sons ever spoke about their father or the divorce. She said Clarence seemed to accept it better than Myers, who swore off all contact with his father as an adult. Myers was always more vocal about his feelings, while Clarence internalized them.

M. C. Thomas died in 2002, just before his seventy-second birthday. Clarence Thomas attended the funeral but felt no emotion, according to his friend Harlan Crow.

When Clarence Thomas was nominated to the Supreme Court many years later, Pin Point became his calling card to the American people. The community's hardscrabble image cloaked him in the American dream and made his story compelling to Americans everywhere. The public never learned of the emotional turmoil that Pin Point represented for Thomas. The splintering of his family scarred Thomas deeply, beginning a pattern of loss and abandonment that repeated itself again and again.

Clarence was quiet and shy as a child, his mother said. Words and pictures fascinated him from an early age. Their clapboard house was plastered with newsprint for protection from the drafts blowing through the slats. As a toddler, Clarence spent hours sitting on the floor staring at the newsprint, especially the funny pages.

As a boy, Clarence palled around Pin Point with Abraham Famble and Isaac Martin, and the three friends ran barefoot around the sandy marshes, catching crabs and making up games. In the hot summers they dove for shade under the wood-frame houses, which were built up on blocks.

In 1954 Clarence started the first grade at Haven Homes, the segregated school his mother had attended as a child. The school was overcrowded, so children attended either in the morning or the afternoon.

"He used to love to get in the house and read, and we'd be playing outside," recalled Martin. "We tried to get him to come out and play marbles, and he'd always have something he wanted to do different."

Clarence became homeless midway through his first-grade year when he returned home to find Aunt Annie's house in ashes. Myers and another boy had accidentally lit the house on fire while trying to ignite the small stove used for heating. Eight-year-old Emma stayed with Aunt Annie in another house in Pin Point. But Clarence and Myers, then only six and five, went to live with their mother in Savannah, boarding with her in a dilapidated tenement with no indoor plumbing. Residents used outhouses in back.

The building fronted a dirt road, and Leola's single room was upstairs on the second floor. One window looked out to the street below.

Leola enrolled Clarence in the Florance Street school on the east side of town. This school, too, was crowded, with students attending only half the day. The rest of the day Clarence roamed the streets or sat at home looking out the window, isolated and alone. Those months seemed "like an eternity," Thomas later said. "It was just bad."

When Leola realized the situation wasn't working for two small boys, she found a slightly better-kept room just a few blocks from her father's house on Savannah's east side. Finally, she could no longer manage the strain and responsibility of working and caring for her two boys. The breaking point came when Clarence was out of school the summer of 1955. "I couldn't afford a baby-sitter," Leola said later. "I didn't want my kids on the street." She went to her father. Would he take the boys during the week, while she worked?

Anderson refused.

Myers's wife, Tena, went into the bedroom and emerged with Anderson's suitcase. "You better pack your clothes," she told her husband. "These are my grandchildren, and I plan to take them."

CHAPTER 4

BLACK AND WHITE

In the summer of 1955, two months after taking charge of Myers and Clarence Thomas, Myers Anderson tried to enroll his grandsons at Saint Benedict the Moor School, across the street from his Catholic church. Nuns accepted young Myers to the first grade but rejected Clarence for the second, finding his first-grade transcript so full of absences that they doubted he had learned enough to begin the next grade. Anderson pleaded with the nuns to make an exception, but the first day of school arrived without Clarence's name on the register of second-graders. Anderson felt he had no choice but to enroll Thomas in the public school he'd attended the year before. For the next forty-eight hours Clarence Thomas's future hung in the balance. On the third day of school, the nuns heeded Anderson's pleas and admitted Thomas to the second grade.

Clarence spent the next nine years in Savannah's segregated Catholic schools, under the stern eye of white nuns and priests. The schools offered the best education possible for a black child in Georgia and equipped Thomas to thrive in successively more difficult school environments. More fundamentally, perhaps, they became the filter through which a young Clarence Thomas experienced race and class, the two external forces that molded his evolving worldview.

Thomas's march through adolescence coincided with the quickening of the civil rights movement in Savannah and the South. For three explosive years, 1960 to 1963, Savannah erupted in sit-ins, boycotts, and protest marches. Time and again the city teetered on the brink of bloodshed, only to turn back. Myers Anderson became a key player in the struggle, using his property holdings to bail out jailed demonstrators, and he regularly brought his grandsons to meetings sponsored by the NAACP.

But the Catholic church insulated Thomas from much of the trauma that other black southerners encountered in their day-to-day dealings with white people and isolated him from the civil rights activism of many black Baptist churches. The Catholic churches of Savannah were never as deeply invested in the civil rights movement as the black Baptist churches for the same reason that made them unique: They were governed and ministered by whites.

Saint Benedict's, a three-story brick fortress about eight blocks from Anderson's house, represented the outer edge of Clarence Thomas's tiny universe in Savannah. Nuns from the order of Saint Francis and priests from the Society of African Missions staffed Saint Benedict's and Savannah's two other Catholic elementary schools for black children.

Dressed in the dark wool habits of their order, the nuns were an intimidating force. The first lesson they taught was how to address them: Yes, Sister. No, Sister. Please, Sister. Thank you, Sister. They demanded perfect behavior from their students. Children were expected to sit upright, raise their hand when they had a question, and speak only when called upon. Children were graded on "deportment," "courtesy," and "application." Penmanship was its own course, and children were graded for neatness. The nuns taught by rote, assigning children to memorize spelling words, addition tables, and Shakespearean verses.

Charles Bell, a classmate of Thomas's brother, Myers, can still visualize the banner over the blackboard in one classroom at Saint Benedict's: IF YOU WANT TO KILL TIME, YOU WORK IT TO DEATH.

The sisters prowled their classrooms and the school halls with rulers, ready to rap the knuckles of any misbehaving student. Sterner discipline included more severe forms of corporal punishment. "You did not have the freedom to do what you wanted to do," said Pamela Jones, who attended another of the Catholic church's segregated schools. "You had to control yourself."

That first year, the nuns' unyielding discipline almost overwhelmed Clarence. Living in Pin Point and then with his mother, he had been mostly unsupervised, accustomed to roaming the streets on his own, even when he was supposed to be in school. His first year at Saint Benedict's, he earned straight C's except for a B in Christian doctrine.

Anderson's towering expectations also pressed on him daily. As much as he believed in the importance of manual skills, Anderson was determined that his grandsons not repeat his life of hard labor. He often told Clarence that he wanted him to study hard so he could land a "coat-and-tie job." Knowledge was the equalizing force in a society that Anderson's generation knew was far from equal. Anderson sometimes put a finger to his head and said that once an idea got in here, no one could remove it. Knowledge was the one freedom white people couldn't take away.

"My grandmother did not have an education, but she demanded that we get an education, her grandchildren, so we could achieve freedom," said Thomas Hines, a classmate of Thomas's brother. "She didn't even know what an education would do, but she knew that she wasn't as free as she wanted to be and we needed to be freer. . . . So education meant freedom."

Anderson expected homework to be completed before bedtime and he tolerated no excuses for staying home from school. He didn't hesitate to remind them of the twenty-dollar annual tuition he was paying to send each of them there. He told the boys that if they died, he would continue to take them to school for three days just to be sure they weren't faking.

Anderson made good on the threat. After missing the first two days of second grade, Clarence was never absent or tardy the seven years he attended Saint Benedict's, according to his school transcript.

From the beginning, Clarence was shy and mostly quiet in school, especially around girls, his classmates recalled. He wasn't the smartest kid in his class, but he was undeniably the hardest worker. There was a discipline to Clarence that set him apart. "He was always about business," said Viola Scott. "I think he was focused from day one."

In class Clarence listened more than he talked. When he asked questions, he was less interested in the short answer than in the fuller explanation of why something was so. "He always wanted to know a little bit more than the average student," recalled Amanda Moore, a classmate. "He wanted to know why something happened, not just that it had happened."

Clarence's cousin Cyrus Green, Jr., recalled that he could rarely convince Clarence to come outside because he was always studying or reading. "If he wasn't doing his homework, he was reading a book," recalled Green. "He always stayed in the books."

By the third grade, Clarence was a solid B student, and in his final, eighth-grade year at Saint Benedict's he earned nothing but A's.

Between his schoolwork and his chores, Thomas had virtually no time for play, even on weekends. Most of his childhood friends remember him in the classroom, not on the playground. "Clarence could never come and play," said classmate Edison Bertrand. "He had to work."

Friends said the cumulative weight of the expectations on his shoulders made him much more solemn than other kids his age. "Clarence had a serious look on his face all the time," said Susan Heyward Williams, a neighbor and classmate. "He was old before his time."

Clarence not only tolerated the nuns' discipline but seemed to embrace it. He became an altar boy at church and assisted with the sacraments both there and at the convent. The duties required Thomas to memorize all the Latin phrases that the priests used to conduct mass, so he knew exactly when to light the candles or assist the priest. Thomas threw himself into the training, proving his commitment by getting to Saint Benedict's before 7 A.M. to assist with the morning mass.

To the best of anyone's recollection, Clarence experienced the nuns' discipline only once. Thomas, Myers, and another boy climbed onto the fire escape at school and dropped a sandbag on the head of a passing priest. None of the boys admitted to the prank, so all three were punished, according to their mother.

Apart from those hours when Anderson permitted him to go to the library, Clarence's only escape was the outdoor basketball court at the school. Basketball was his first love, although he also played football and baseball. On the blacktop, his buddies saw in Clarence a competitive drive and showmanship not evident in the classroom. Thomas was nicknamed "Cousy" or "Couze," after the Boston Celtics' Bob Cousy. Cousy was known as one of the best ball handlers of his day, and Clarence got his name because he was a good dribbler.

What Joe Williams remembers most about Couze, however, was that when he got the ball he was going to show his teammates that he could dribble. Clarence didn't just dribble the ball, he bounced it between his legs and around his back. He twirled. He spun. He hogged the ball so much that his teammates gave him another name: "the Black Hole." Give the ball to Couze, and it was swallowed into the Black Hole. Anderson didn't allow Clarence to play basketball on the weekend if there was work to be done, so he primarily played during school recess. Perhaps he hung on to the ball so long because he just didn't have the chance to play with it all that much.

· · ·

Clarence's studiousness and focus, so clearly recalled by classmates, in fact concealed a swirl of raw emotions his first years in Savannah. Living in the city and attending Catholic school exposed him to racial and class tensions that he'd not felt in Pin Point. Suddenly, he saw that his world wasn't simply white and black but a hundred shades of gray. He saw how wealth and status organized Savannah's brutal pecking order in the black community.

Anderson, an unlettered tradesman, orbited in the lower spheres of Savannah's caste system, but even he wasn't immune to the elitism that infected black life, derisively referring to Pin Point as "dem people."

"There were a few places where if you were from, you were just at the bottom," recalled Thomas. "And we were from Pin Point. We were 'dem people out dere.'"

Clarence's classmates were mostly the sons and daughters of Savannah's black upper and middle classes, the children of doctors, dentists, teachers, and ministers. The class distinctions were also often tied to skin color, with fair-skinned blacks at the top of the heap. The differences determined who your friends were, where you lived, and whom you could date.

"You had the college-educated blacks, and they were in a different class," recalled Edison Bertrand, whose father was a professor at Savannah State College. "They lived in better neighborhoods. And they had social clubs that did not include those who did not have those benefits." Bertrand's mother belonged to Jack and Jill, a social club that met regularly to celebrate birthdays and other special events. "There were no dark people in Jack and Jill," said Bertrand. "The closer you were to being white, the better you were. Being white meant being free. It meant having opportunities. It meant being favored. At that time, what was there to be favored about being dark?"

In the 1950s, before the days of black power and "Black is beautiful," being called "black" was an insult, recalled Earl Calloway, who played basketball with Thomas. "Those were fighting words," said Calloway. "No one wanted to be black. Skin tone meant a lot."

The class and color distinctions manifested themselves cruelly among schoolchildren. A contemporary of Thomas's, Sage Brown, recalled being invited to a birthday party by the prettiest girl in his grade-school class. Later the same day the girl's mother, a teacher at the school, quietly but firmly demanded the invitation back. Brown, dark-skinned and from a poor family, he learned he wasn't welcome in certain black homes. "It was horrible," Brown recalled. "It was almost like white folk never got a chance to put me down because black folk would do it to me so much."

The distinctions extended beyond skin color. Straight hair was better than kinky hair. Delicate bone structure was better than broad. "The kids who had straight hair and light skin were treated differently—even in the Catholic schools," recalled Pamela Jones. "The discipline was harsher for people of color. There was a distinction."

In appearance, Thomas favored both his mother and his grandfather. He had Anderson's broad nose and coal black skin. His hair was curly, not flat, and his lips were wide, not thin. From birth, Thomas was pigeon-toed, and he walked with an open gait that made him look slightly bowlegged. Thomas was "no Denzel Washington," as Earl Calloway recalled. In the competitive, class-conscious world of young kids, all of Thomas's physical attributes became easy targets, especially for the girls. "Clarence was looked over a lot because he was dark-skinned," said Viola Scott. "He just didn't fit in."

In the school yard Thomas earned the nickname "ABC: America's Blackest Child." "The reality was the more Negroid you were the more you got it," said Thomas. "So me, I'm Negroid, I took my hits."

According to his classmate Robert DeShay, Thomas was also ridiculed for his Pin Point dialect, which immediately broadcast to others that his parents were not educated, and teased for being a "Geechee." "Geechee was really a different form of 'nigger,'" said DeShay.

Thomas downplays the seriousness of the racially motivated teasing, but his mother said it stung him deeply as a child. "Sometimes, I wonder how some of the anger hasn't come out of him because of the names they called him," she said.

Thomas confided the extent of the teasing only to his brother, according to his mother. There was simply no one else to talk to about it. "This was something you didn't talk about with anybody," said DeShay. "You were ashamed of who you were because of the color. And you were more ashamed that so many people were beating you up for it."

Anderson, despite his business success, was never a part of Savannah's elite class of black businessmen, and he did not belong to the lodges and fraternal societies where upper-class black men socialized. Indeed, he suffered almost as many insults from elite blacks as from bigoted whites.

Anderson never forgot a particular visit to Savannah State College to inquire about courses for Thomas's sister, Emma Mae. Administrators sat him in a waiting room and then ignored him as though he was a homeless man off the street. Anderson was livid, recalled his daughter.

"You had the black elite, the schoolteachers, the light-skinned people, the dentists, the doctors," said Thomas. "My grandfather was down at the bottom, an uneducated man who had money in the bank and took care of himself. And... they would look down on him. Everybody tries to gloss over that now, but it was the reality. It was the reality."

Thomas deeply resented the condescension other blacks showed him and his grandfather. The treatment became the germ of a class-consciousness that remains with him to this day, and the prism through which he ultimately viewed many of the attacks waged against him as an adult. As a child, Thomas learned that both races could be equally cruel. "People love to talk about conflicts interracially," said Thomas. "They never talk about the conflicts and tensions intraracially. The ones we had to deal with most often and most frequently were the intraracial ones. And it gives you a different perspective on life and the things that are happening to you."

The enforced separation of blacks and whites in Savannah added yet another dimension to the racial turmoil of Thomas's early life. Segregation appeared to a young Clarence Thomas as a series of invisible stop signs, sharply confining where he traveled and whom he played with. Savannah's schools were the most obvious manifestation of segregation, but they were by no means the only one.

Weekend afternoons, Thomas and other Catholic school kids loaded into a church bus to make the three-hour ride to Fernandina Beach, Florida, and nearby American Beach, as it was called, the colored beach. They swam there because not a yard of Georgia beach was open to black bathers. "The water was foul," recalled Patricia Luke, one of Thomas's classmates. Thomas never learned how to swim; a near-drowning on one trip made him fearful of the water to this day.

Apart from these trips, Thomas did not set foot beyond Liberty County until he graduated from high school. He lived his life in three Georgia counties—Chatham, Liberty, and Bryan, the county in between. Anderson dared not take his family anywhere else during segregation because he dared not risk exposing them to potentially hostile situations. He traveled only where he knew how to navigate safely, nowhere further.

In Savannah Thomas lived most of his childhood in an area that was ten blocks square. He learned to avoid neighborhoods where he might face hostile white children. He did not ride the segregated public buses or shop downtown, where blacks routinely received second-class service.

At the segregated high school Thomas attended for two years, the only competitive sport the school could field was basketball. With so few black schools, there simply weren't enough teams to play. Under segregation, games between white and black teams were forbidden.

The rules all but eliminated contact with whites.

"You might have some occasional contact [with whites], but your world was black and you were supposed to stay within that world," said Thomas. "You might see whites drive by, or on TV, but you just didn't deal with them."

Youthful imagination filled in the gaps. In Liberty County, where he spent the summer, Thomas knew of a local white judge who lived in a house near the water. Not daring to venture close, he pictured the man living in the splendor of a plantation house. Years later he met the man and saw the house: It was tiny.

In Clarence's household, Anderson was polite to whites, his grandson recalled, but he never, ever trusted them. Anderson talked about whites in the coded language his slave ancestors used to describe their owners. He called them *buckra*, a West African word for "demon" that shows up frequently in the slave narratives recorded by the federal government in the 1930s. "You don't want to be messin' with no *buckra*," he repeated, teaching his grandsons to avoid racial conflicts.

Anderson stared down virtually every stigma segregation pushed on him, but he could never walk into City Hall to obtain his annual business permit without a shot of corn liquor first. A stiff drink was the way Anderson coped with the degrading experience of waiting to be attended to, or listening to a smart-ass bureaucrat call him "boy."

Years of such experiences taught Anderson and other blacks of his era to believe that government, like many other things in the segregated South, was for whites only. "My grandfather told me that when he was raising his kids during the Depression, there was no New Deal," recalled DeShay, whose grandfather was a contemporary of Anderson's. "He said, 'None of that stuff came to us black folk.' It was, 'I had to do, and I had to, and I had to.' My grandmother said, 'We couldn't depend on the police. We couldn't depend on the politicians. We couldn't depend on anybody. The only faith we could put out there was our faith in God.'"

Segregation was the reason Anderson was in business for himself. He had worked for whites in his youth, and he never wanted to work for them again. Segregation was also the reason he worked Thomas and his brother so hard. He wanted his grandsons to understand they could not depend on white people for help. No white bank lent Anderson the money he needed to start his business or build his own home; he saved it himself. In a world that favored whiteness, a black man had to work twice as hard to go half as far. For a black man, hard work was the essence of survival.

Segregation insulted Anderson's sense of fairness and his beliefs about the intrinsic equality of whites and blacks in the eyes of God. Although Anderson's eternal optimism prevented him from complaining too openly about segregation, he loathed all that it represented. "He knew it was wrong that we couldn't eat at the counters downtown, he knew it was wrong that we had to sit on the back of the bus," said his friend Sam Williams. "Those things just tore him to pieces."

For Clarence Thomas and other black Catholic children, however, segregation was different in one crucial respect: They went to school every day with white nuns. Not only were the nuns and priests of the Catholic schools all white, they were mostly all foreign, joining the diocese from Ireland, Scotland, and Wales. Unlike most southern whites steeped in white supremacy and Jim Crow, the Franciscan sisters arrived in Savannah with no demeaning stereotypes of black children. "They were just children to us," said Sister Mary Virgilius Reidy, Thomas's eighth-grade teacher.

"They never cared what the color of our skin was," said Dr. Charles Elmore, a Savannah historian and product of the city's Catholic schools. "They were concerned about you as a person and you being all you could be."

The nuns tolerated but never embraced segregation, burnishing their image outside the classroom with a few well-chosen acts of defiance. Ethel Heyward, whose fourteen children all attended Saint Benedict's, recalled how two sisters escorted a small group of black children downtown to go shopping in the late 1950s and were declined service at a department store. The nuns refused to leave until they—and the children—were given food. "I don't know what the sisters said, but after they got through with them, they got served," recalled Heyward. Another sister liked to walk her class to nearby Forsyth Park but was repeatedly told the children weren't to use the public bathrooms. "Okay, then we'll use your trees," she responded, according to Sister Carmine Ryan. During a basketball game with the white Catholic school in the 1960s, one nun became convinced the white referee was calling an unfair game against the black players. She marched down onto the gym floor at halftime, stuck her nose in the referee's face, and accused him of bigotry, recalled Orion Douglass, who played on the black team.

The sisters became so closely identified with the black children they taught that they ceased being "white." White Savannahians called them the "nigger nuns." For whites, the nuns were as black as the children they taught, and both drew equal contempt.

Thomas and his peers remember the nuns' strict discipline, but what they remember most was that these white women, sworn to vows of poverty, devoted their whole lives to teaching them how to learn.

The nuns made it clear that they expected their pupils to perform as well as—or better than—any white child. Like Anderson, they believed that failing was never an option. "You knew you were there to learn," said William Fonvielle, one of Thomas's contemporaries. "There was no question why you were at school."

To this day, Lester Johnson, a Savannah attorney who followed Thomas at Saint Benedict's, can vividly recall how the nuns there admonished him and his classmates: "'I know you are smart, so don't come in here telling me you don't know the answer to that question. I know better,'" said Johnson. "That had to have a subconscious impact on how we felt about ourselves.... Nobody could ever tell us that we weren't smart."

"We knew that segregation and discrimination existed," recalled

Joseph Williams, who played basketball with Thomas. "But those nuns instilled in us that we were equal."

More subtly, the nuns taught their students that it was okay to be different. As black Catholics, Clarence Thomas and his schoolmates were part of a tiny minority within Savannah's black community. When Clarence Thomas was baptized into the church in the 1950s, there were fewer than twenty-five hundred black Catholics in the entire Savannah diocese, which covered one third of the state of Georgia.

As a black Catholic in an environment that was Protestant and white, Thomas learned to think of himself as someone different and apart from everyone else. The church gave him the courage to hold contrary beliefs—the character trait that defines him today. "I don't need people to agree with me," he told a group of college students a decade after ascending to the Supreme Court. "I'm very comfortable being alone in my views."[1]

The nuns also made their students feel differently about white people. Students at Saint Benedict's learned that not all white people hated them. The nuns' compassion resonated doubly with Thomas because of the scorn he experienced from kids of his own race. "So many of the people who attack Clarence fail to understand that he was taught by white nuns," said Edison Bertrand. "These were the people who helped him. These were the people who encouraged him."

The nuns never hugged or showed affection to their students, but Thomas has been moved to tears talking about the sisters of Saint Benedict's.

Years after he left Savannah, Clarence Thomas as a young man looked up Sister Mary Virgilius Reidy, his eighth-grade teacher, who was then living in Boston. He wanted to know only one thing, Sister Virgilius recalled: "Why was it that you sisters could do for us black kids what nobody else could or did do?"

. . .

On March 16, 1960, two black teenagers and a local black college student walked into Savannah's biggest department store and asked for something to eat in the restaurant. The store manager gave them two minutes to leave. When they refused, he called the police and had them arrested for trespassing. The arrests, coming weeks after student sit-ins in Greensboro, North Carolina, ignited the modern-day civil rights movement in Savannah.

As an eleven-year-old in the sixth grade, Thomas was too young to participate. His opportunity for protest came later, when he was in college. Yet the tumult of the Savannah civil rights struggle played a key role in shaping his beliefs about equality and discrimination. Even as an adolescent he picked his own battles to make his voice heard.

The Savannah chapter of the NAACP, headed by Anderson's friend, W. W. Law, was the driving force behind the movement. Anderson had joined the organization in the 1940s, after a crusading Baptist minister named Ralph Mark Gilbert had resurrected it from oblivion. Although never active in the NAACP's leadership, Anderson was a faithful financial contributor during the lean years of the forties and fifties.

After the outbreak of protests, Law held weekly "mass meetings" to communicate the aims and needs of the struggle. He borrowed the idea from Dr. Martin Luther King, Jr., who organized mass meetings during the Montgomery bus boycott. The meetings were held at four o'clock every Sunday in a different Savannah church. Anderson attended regularly, and he often brought both his grandsons.

That he brought his grandchildren to the meetings was in itself unusual. Many parents sought to shield their children from the protests, fearful that they, too, might be swept into jail. Protests and demonstrations, so common in America today, were fundamentally and radically contrary to how black southerners had endured segregation. Black families were used to avoiding confrontations with whites. Unlike Anderson, who worked for himself, most black men and women owed their employment to whites and risked their jobs by openly de-

fying the status quo. The trepidation many felt spread to how they
guided their children through the unrest of the civil rights movement.
In Thomas's neighborhood most families forbade their children from
participating.

"I was never involved in any of the civil rights demonstrations
when I was living in Savannah," said Robert DeShay, Thomas's boyhood friend. "That was not allowed in my family. Most of us in
Catholic school, we came from very conservative, nonconfrontational
families. The approach was to overwhelm them with competence,
overwhelm them with excellence, batten up your hatches with credentials, knowledge, and information. But you do not confront these
powers in this kind of way."

The Catholic church provided an extra layer of insulation. The
Sunday meetings almost always took place in the Baptist churches.
Though generally supportive of civil rights, the Catholic churches
were not actively involved in the day-to-day planning of the movement. Thomas and other black Catholics didn't hear the weekly drum-
beat of activism that was so vocally articulated in the Baptist churches
as the movement gathered force.

Thomas absorbed the civil rights movement by reading what little
coverage appeared in the local newspapers and by listening to the sermons preached at the mass meetings. Week after week, the speakers
sought to reinforce the idea that it was not just reasonable but morally
imperative for black people to stand up and fight a system that had
been used for so long to keep them down.

Many speakers spoke of the "new Negro" who had been born out
of the gathering civil rights movement. Negro youth of the 1950s,
said teenager Jesse Blackshear, one mass-meeting speaker, were only
interested in "clothes, rock and roll music, and material things in life."

"But the new Negro youth of the 1960s is not interested in the
frivolous and the ephemeral," said Blackshear. "No longer will Negro
youth accept segregated lunch counters, rest rooms, and water foun-
tains."[2]

As a boy on the cusp of manhood, Clarence Thomas straddled the dividing line between the old and the new Negro of his day, with one foot planted in the 1950s of his childhood and the other stretching out precariously toward the unknown world beyond. Thomas recalls the time as a sudden "rush to consciousness about race" that was almost overwhelming. The civil rights movement forced to the surface ugly realities that decades of segregation had tried so vainly to conceal. Suddenly, it was plain to everyone, including Thomas, that segregation was a system rooted in racial hatred. White southerners wanted to preserve segregation because they simply didn't want to live with blacks on a basis of equality. The realization disgusted Thomas. "The longer you talked about it and thought about it, the worse you felt about it," he said. "And the angrier you got."

The often tense atmosphere in Savannah during the civil rights protests affected him keenly. At times he imagined he was walking through dense fog on a wooden bridge that was slowly being consumed by fire. "There was this nothing behind," he said. "So, I had to keep going forward."[3]

He sought solace in the only place of safety that he knew: the Catholic church. Thomas threw himself into his faith with renewed fervor during those tumultuous years. In 1961, when he was in eighth grade, Thomas was named altar boy of the year, a prize reserved for only the most devout and committed. The accolade started him thinking about the priesthood as a vocation. In a world consumed with hate, God's love and protection seemed a safe haven.

Like so many of his generation, Thomas was drawn to the passion and idealism of the Reverend Martin Luther King, Jr. Anderson kept a framed picture of the Baptist minister at home, and Thomas read his speeches and sermons. King's appeals to universal human dignity resonated profoundly with Thomas, echoing the church's teachings about the equality of all people before God.

King's plea for all Americans to be regarded for the content of their character, rather than the color of their skin, doubly resonated with

Thomas because he felt so strongly about being judged and catego-
rized as a dark-skinned man. King's ideal formed the kernel of a
color-blind worldview that Thomas embraced the rest of his life.

As a teenager, Thomas registered his opposition to segregation with
a few acts of defiance. In the fall of 1961 Mayor Malcolm Maclean
declared that Savannah's libraries would no longer bar blacks. Thomas
stopped going to the black-only Carnegie Library at night and instead
went to Savannah's main public library on Bull Street. He sat in the
airy reading room, previously reserved for whites. White readers stared
at him, uncomfortable with his presence. Thomas stared back, as if to
say, "Get used to me." Eventually, they did. Today Thomas stares down
his critics with the same stubborn defiance.

As an adolescent, Thomas longed to go to Savannah's Weiss Theater and see the James Bond movies that were then the rage. But he refused to buy a ticket because he would not sit in the balcony and be
made to feel inferior. To this day, he has stubbornly refused to watch
James Bond movies because he associates them with the segregated facilities of his youth.

Thomas graduated from Saint Benedict's in 1962 and entered the
all-black Catholic high school that fall. Just as strict and demanding,
Saint Pius X also drilled into Thomas the church's taboos about sex
and abortion. He and his classmates were taught that God made sex
so that men and women could reproduce, that sex before marriage
was a sin, and that abortion was immoral. "Sex was a bad thing," recalled Viola Scott. "It just wasn't done unless you were married, and
then only to have children."

Vern Cameron said the nuns' preaching about abortion has stayed
with him his entire life. "It's murder. It's wrong," said Cameron,
whose mother was the Carnegie librarian. "That was taught in school.
There wasn't any prettying it up. It's just wrong to kill a baby. It's
wrong to have sex before marriage. It's a sin."

Thomas internalized the church's discipline and embraced its abso-

lutism. Thomas has never been a moral relativist. From an early age, he learned to see the world in sharply defined categories of good and bad, heaven and hell, virtue and sin. This way of evaluating the world one day added up to a judicial philosophy that was equally categorical and absolute.

In the fall of 1963, when Thomas was a sophomore, city officials ended segregation in most restaurants, movies, and public places. The announcement followed weeks of tense standoffs between police and demonstrators; at its peak some three thousand people marched through downtown Savannah streets.

On New Year's Day 1964 Martin Luther King visited Savannah for the second time since 1961 and declared it the most integrated city south of the Mason-Dixon line.

Six months later President Johnson signed into law the 1964 Civil Rights Act, ending nearly one hundred years of segregation in the South. Title VII of the new bill, banning employment discrimination, created a new federal agency to combat workplace bias. The new agency was called the Equal Employment Opportunity Commission, and Clarence Thomas was one day to become its longest-serving chairman.

Segregation ended when Thomas was sixteen years old but left a permanent legacy on his life. It instilled in him a visceral loathing of classifying Americans by race, the foundation of his opposition to affirmative action, and made him determined to live his own life unbound by racial assumptions. For Thomas, segregation's lesson was clear: Using race to gain an advantage was morally and legally wrong. Today Thomas sees affirmative action as the same principle at work for different ends. And in Thomas's absolute worldview, ends never justify means.

Segregation also became the springboard for Thomas's libertarian political philosophy and the basis for his deep, continuing mistrust of government. Thomas saw segregation as the ultimate abuse of govern-

mental power, a dishonest system in which white people twisted public laws to benefit themselves. He knew segregation was the reason
Anderson never finished school, the reason he was an oil deliveryman
and not an oil distributor.

Thomas deeply resented segregation's double standards. Later, the
sensitivity translated into a rigid determination to judge and interpret
laws so they treated all Americans equally, regardless of color or social
status. "I lived under two sets of books," he said nearly a decade after
ascending to the high court. "I'm not going back to two sets of books
again."[4]

In Savannah, segregation's demise impacted Thomas's life almost
immediately. In 1964 the Catholic diocese moved ahead with plans to
integrate an elite boarding school for aspiring priests outside of Savannah. Church officials visited Saint Pius's and talked up the seminary school to prospective black students. Thomas, a tenth-grader,
heard the pitch and volunteered to transfer to the all-white school.

Over a game of hoops one day, he casually told his Saint Pius basketball buddies that he planned to start at the new school in the fall.
His friends were stunned. "We were out there chasing girls and
stuff," recalled Earl Calloway. "We wanted to date. The idea of becoming a priest was totally foreign to us." Thomas continued to play
basketball with the guys some that summer. Then one day, Calloway
said, "he just disappeared."

CHAPTER 5

THE ISLE OF NO HOPE

In the fall of 1964 Myers Anderson deposited his grandson at the steps of Saint John Vianney Minor Seminary. Fewer than six miles separated Anderson's home in Savannah from Saint John's, but the distance could easily have been six hundred miles. The school sat forlornly at the end of Grimball's Point, a deserted spit of sand that backed up to Grimball Creek, one of the many small waterways feeding into the Skidaway River. Grimball's Point was actually part of an island. Savannah's early settlers named it the Isle of Hope. Boys at Saint John's gave it a different moniker: "the Isle of No Hope."

At sixteen, Thomas should have come to Saint John's as a junior. But school administrators held him back, telling him he needed an extra year of Latin to complete graduation requirements. Just five years old, Saint John's was the most elite, challenging, and rigorous Catholic school in south Georgia. For Thomas it also held the promise that he could become Savannah's first black Catholic priest. The dream all but consumed his grandfather, who prayed he might live to see his grandson crowned with the church's highest accolade.

Saint John's also appealed to Thomas because it was a chance at independence after ten years of living under Anderson's iron rule. His confrontations with his grandfather had become increasingly sharp after he started high school, with basketball a primary bone of contention. Thomas wanted to play. Anderson wanted him on the oil truck. On another level, however, the teenage Thomas was simply becoming more assertive. Anderson started calling his grandson "Mr. Know-it-all," a dig at Thomas's increasing boldness.

Anderson gave Thomas one piece of parting advice: "I done all I can for you. Now, it's up to you."

Yet as he lay awake that first night in the sophomore dormitory, a low-slung barracks charitably described as spartan, Thomas began to comprehend the reality of his new life. Twenty-two boys bunked down in their beds that night. Twenty-one of them were white. Thomas had never been so acutely conscious of his skin color or felt so alone.

Nighttime in the dorm seemed to bring out the worst in his new classmates, Thomas learned.

"What's that smell coming from the other end of the room?" they called out under the cover of darkness. "It's so dark, I can't see somebody."

"Smile, Clarence, so we can see you."

Across the hall in the freshman dorm, Richard Chisolm, the school's only other black student, endured the same taunts. Daytime was slightly better, recalled Chisolm. Then the white boys "just ignored you."

"You felt all alone," Chisolm recalled. "There was no one there but you."

The next ten years were some of the most difficult and tumultuous of Thomas's life. After Saint John's, Thomas attended a Catholic seminary in Missouri and the College of the Holy Cross in Massachusetts. In 1974, a decade after he began at Saint John's, he graduated from Yale Law School. Over the same decade, America fought its longest war, cities exploded in racial bloodshed, and college campuses became staging grounds for radical protest.

During those years, Thomas traded the cassocks and surplices of the priesthood for the combat fatigues and berets of the black nationalists, veering sharply to the left in his political thinking. At Yale, the army fatigues were replaced by bib overalls and knit caps, badges of Thomas's working-class roots among Yale's ivy-covered towers. The blue-collar clothes ultimately gave way to the conservative business suits of law practice.

The journey through high school and college tested Thomas's strength and discipline at every turn. There were moments when he believed he did not have the will to continue, and he thought many times about giving up. The first year at Saint John's, Thomas recalled later, was the most difficult.

Thomas stared down bigotry there and learned to stop listening to the voices that whispered, *You don't belong here. You aren't good enough.* He learned to believe in himself at Saint John's, taking his first steps toward becoming his own man. Thomas has always marched to his own drummer. He heard its first syncopated rhythms in a tiny community of teenage Catholics on the Isle of No Hope.

The campus where Thomas spent his next three years was actually built as Camp Villa Marie, a summer retreat for Catholic families in the 1930s. The centerpiece was a small chapel that formed one end of a three-sided courtyard. Two single-story classroom buildings formed the other sides of the enclosure. In the center of the small courtyard stood a tall, slender statue of Saint John Vianney, eyes downcast and hands pressed together in prayer. The dormitories sat on the far side of the basketball court and playing field, just a hundred yards from Grimball Creek.

The prep school had opened in September 1959. Four priests provided most of the instruction, with a few nuns and other priests pinch-hitting in specialized science classes. Two black women cooked the meals and tended to the kitchen, and a black caretaker served as maintenance man, groundskeeper, and custodian.

Religious indoctrination permeated school life. Students began their day with meditation and prayers, followed by morning mass. Afternoon rosary preceded lunch. More prayers and spiritual readings preceded bedtime. During high holidays such as Good Friday and Easter, seminarians spent up to four hours praying in the chapel—all on their knees. Inside the classroom the students studied religion and Latin. They engaged in endless debates about the meaning of faith, the will of God, and the nature of sin.

For Clarence Thomas and every other seminarian at the school, Father William Coleman, the rector and dean, personified the priestly rigors of their life. He taught both Latin and religion, the two courses that defined the church's core theology. No other figure commanded more respect or fear. Coleman ruled Saint John's with the discipline of a Marine Corps drill sergeant and the religious fervor of a monastic.

A native of Connecticut, Coleman was determined to fashion Saint John's into the most demanding and God-fearing seminary in the Southeast. His wide eyes burned with intensity. He wore a confident, almost cocky expression that spoke to the strength of his convictions and the power of his intellect. He was solidly built in a way that suggested he was no priestly bookworm. Although he was not yet thirty, his hair was prematurely gray, an attribute that only enhanced the authority that he exuded in his crisp commands and sometimes sharp rebukes.

Coleman imbued his religious beliefs with a fierce sense of social responsibility. He believed that he was educating his seminarians not just to be devout Catholics but also to be evangelists for social justice. Over the summer, when he wasn't teaching, he volunteered at an orphanage run by the diocese.

"We hope that each boy who comes to us will grow in his ability to love others, to guide them and to assume responsibility for the direction of tomorrow's world," Coleman said in 1964, the year Thomas arrived at Saint John's. "We try to teach here that those who really count in life are not the men who make fortunes and live comfortably, but those who assume the reign of leadership and influence future generations."[1]

Coleman barely concealed his contempt for some of the Catholic church's worldly excesses. Students recalled him snorting loudly when church officials arrived in their expensive Buick sedans for periodic school inspections. Another example of "good old Christian piety," he muttered sarcastically. Coleman complained that such trappings of wealth were unseemly for followers of Jesus, whose own life he believed defined God's humility.

Coleman's daily Latin class was a lesson in fear. He taught by recitation and translation, and he expected his students to be prepared. Lateness resulted in automatic demerits. To this day, Mark Everson, one of Thomas's classmates, is stricken with anxiety over the thought of being late, a fear he traces to being late to Coleman's Latin class. For three years Everson preceded every Latin class with a furtive stop in the chapel to say a prayer of help for the peril that lay ahead. After each class he returned to the chapel to say a prayer of thanksgiving.

Coleman's effect on his students stemmed from his stratospheric expectations. When called on, a student was expected to stand and translate the homework passage or recite it aloud from memory. Anyone who wasn't prepared received an automatic zero for the lesson, an academic hole that was almost impossible to climb out of. Sometimes even hard work wasn't enough. One of Thomas's classmates scored a perfect 100 on a Latin exam. But when Coleman read the scores aloud, he awarded the student a 99. What happened to the other point? the student wanted to know. It was deducted, Coleman replied, because no one was perfect.

"He had no mercy," recalled Chip Boyett, one of Thomas's classmates. "He pointed out your weaknesses, and he prodded you on to higher ground. At times he could be brutal."

Yet Coleman's toughness also conveyed a deep sense of caring for the academic development of his students. If he showed a soft side, it came through in his tolerance for the endless pranks played on him. Students booby-trapped his office with smoke bombs; they replaced his lightbulbs with colored lights; they put peanut butter in his shower nozzle. Coleman never disciplined the boys for such pranks and even seemed to encourage them. After each attempt to surprise him, he invariably retorted that the prank showed "no originality."

Coleman arrived in the heart of Dixie with strong views about racial prejudice and segregation. His students knew little about his involvement in the Savannah civil rights movement. After class, he often disappeared to attend civil rights meetings, and he became passionate about the cause of integration and the demise of Jim Crow. After Congress passed the historic Civil Rights Act, Coleman insisted that the Savannah diocese integrate Saint John's. Clarence Thomas and Richard Chisolm were the products of his insistence.

At the time, most white southerners believed black children were intellectually and socially inferior to white children. The attitude expressed itself in newspaper editorials, letters to the editor, and court briefs filed to block integration. In May 1963, for example, the Chatham County Board of Education and Georgia attorney general Eugene Cook filed briefs seeking to stop the Fifth U.S. Circuit Court of Appeals from integrating the Savannah public schools. The briefs cited intelligence and achievement tests administered in Savannah showing that only 15 percent of black students equaled or exceeded the median scores of white students, and "mental maturity" tests on which whites scored twenty points higher than black students. "Marked differences remain even after the socio-economic factors are equated, thereby illustrating an 'inherent' as distinguished from environmental origin," the all-white board of education argued.[2] Although the battle for integration was slowly being won, Thomas learned in school that the racist underpinnings of segregation were not so easily expunged.

"The assumption was you were stupid," said Sage Brown, one of the nineteen students to integrate Savannah high schools the year before Thomas started at Saint John's. Before his senior year at Groves

High School, Brown had always loved school. But the trauma of learning under such racist assumptions about his intellect, while enduring daily taunts and ridicule, crushed his love for school and nearly destroyed his psyche. "That whole year was just a nightmare," Brown recalled.

Saint John's held one distinct advantage over Savannah's public schools because it was a religious school filled with boys intending to commit themselves to spiritual devotion. Brown and others knew they were attending school with hostile white children who would sooner spit in their face than welcome them. Thomas and Chisolm were expecting something different, and they got it. But neither Father Coleman nor daily prayer kept out the subtle reminders of racial prejudice, beliefs, and attitudes that had stewed unchecked in the Deep South for more than two hundred years. "It was a racist culture," recalled Everson. "And most of our parents were probably racists. So we had that cooking in our blood."

Acutely aware of white assumptions about the abilities of black children when he entered Saint John's, Thomas became hypersensitive to the slightest suggestion of inferiority. In his eyes the signals were everywhere, beginning with the fact that he'd been held back a year. The school told him it was because of the Latin requirements. Or was it because they thought he wasn't as smart as the white students? What did his classmates assume about his being held back?

In the dining hall Thomas noticed that some boys passed by the empty seat next to his. On the basketball court the custom was to allow the player who sank the first basket to choose up sides. But when Thomas made the first basket, the rule was disregarded. Then there were the nighttime taunts in the dormitory.

"Some of us were real racist jerks to him," said Everson. "Most of us were clueless at the time about what it was like, and how we were really judging and putting down Clarence and his background."

Thomas felt homesick and alone his first few months at Saint John's. He thought about quitting. Yet a part of him was determined to succeed. Anderson and the nuns had taught him he could overcome anything by working hard. He threw himself into his studies and attacked his schoolwork with a vengeance. His diligence paid off. His grades were good—not great—but good enough to make him believe he could succeed.

Toward the end of the first semester, however, Father Coleman summoned Thomas to his office for a private consultation.

Coleman's style in such encounters was blunt and to the point. "He would not preface a remark by holding your hand," recalled Boyett. "It was like a cannon in your face." The dean of students told Thomas that his performance was average and that his English was "atrocious." He told him that if he had any pretense of succeeding at the school, or later in life, he would have to learn to speak "Standard English," not the black dialect of the Gullah/Geechee. Father John Fitzpatrick, another teacher at the school, was with Thomas in Coleman's office. Fitzpatrick shifted uncomfortably in his chair in the ensuing silence, his own indignation rising at what he sensed was an insult to the young black boy.

Thomas sat stunned. The remarks cut deeply. Black children in Savannah had teased him about his dialect. For him, the criticism was tantamount to telling him he was just some ignorant black kid from the fields.

The scar from that day was one Thomas carried for years. Father Fitzpatrick reminded him of the remark thirty years later, after Thomas was on the Supreme Court, when the two shared a quiet car ride together in Augusta, Georgia. "He said to me that made him come to the conclusion that *no one* would ever say anything like that to him again," Fitzpatrick said.

Coleman's criticism made Thomas doubly self-conscious about the way he spoke, and it affected his behavior for years. In college Thomas majored in English, even though other subjects interested him far more, because he perceived it as his weakness. Perhaps more lasting, Coleman's criticism made Thomas feel shy and uncomfortable speaking in public among white people. At Saint John's he became acutely aware of how white people judged him, silently evaluating his intellect and character. He resented that arrogance and decided that sometimes he would simply keep his mouth shut.

"It was hard because it was white," Thomas said of Saint John's years later. "You still had a society that said you weren't supposed to do the same things white people did."

At the conclusion of his conference with Thomas, Coleman told the sophomore seminarian that he planned to tutor him in Standard English, as he called it. The lessons were to be one-on-one sessions in the evenings. Thomas would read aloud. Coleman would correct his articulation and pronunciation. The sessions, Coleman made it clear, were not optional.

Years later Thomas learned that Coleman was equally harsh and demanding with all of the other students. And he came to respect Coleman's honesty, crediting the stern rector with giving him the tools to succeed in white America. But at the time Thomas was convinced Coleman had singled him out because he was black.

His first taste of accomplishment came not in the classroom but on the playing field. Thomas had always been a good athlete, but at Saint John's he became a superstar. His football throwing arm was overpowering; he could literally knock a receiver down. Though short, Thomas was a standout on the basketball court, dazzling the white boys with his dribbling prowess. Whatever his classmates thought of him in the classroom, they learned to respect Thomas on the playing field. His athletic ability gave him confidence that he desperately needed in those first, difficult months at Saint John's. "He felt like when he did sports and he did well, people cheered him on and embraced him as a human being," said Armstrong Williams, who worked for Thomas in the 1980s. "They saw his humanity."

Soon that confidence began to carry over to his schoolwork. If he could beat the white boys on the field, why couldn't he also beat them in the classroom? he began to wonder. The goal of succeeding against the racist expectations of his classmates stoked his competitive drive and made him keep going even when he thought he didn't have the strength to continue.

Thomas won a Latin spelling bee in Coleman's class and was awarded a small statue of Saint Jude, the Catholic patron saint of hope. He put the statue on a small bureau next to his dormitory bunk. "A few days later, I looked over and saw the head was broken off, lying there right next to the body on my bureau, where I'd be sure to see it," Thomas told a journalist years later. "I glued it back on. After another few days, it happened again. So I got more glue—put it on real thick—and fixed it again. Whoever was breaking it must have gotten the message: I'd keep gluing it forever if I needed to. I wasn't giving up."[3]

Black American authors also gave Thomas strength. He read Ralph Ellison's *Invisible Man* and novels by Richard Wright. Often he read at night, after official lights out in the dormitory. With a small flashlight for illumination, he huddled under his covers and devoured *Native Son*, Wright's riveting account of another Thomas, named Bigger, an angry black man who kills a wealthy white woman in a moment of panic. The girl's mother, a blind woman, nearly catches Thomas putting the young woman to bed after a night of heavy drinking with her boyfriend. Bigger Thomas is seized with horror: he, a black man, alone in a bedroom with a white woman. He accidentally suffocates the young woman with a pillow trying to avoid detection. The novel spoke forcefully to the reality Thomas knew as a young black man living in Savannah. "I was just angry," said Thomas. "The more I read, the angrier I got about just the way blacks were being treated. . . . I read the newspapers voraciously. I read magazines. I started reading novels I hadn't read before, literature I hadn't read before.

"He's an angry black novelist," Thomas said of Wright, "and I was an angry black man."

Thomas was so quiet and reserved, however, that none of his classmates sensed what he was feeling. Classmates don't recall Thomas so much as complaining. "I don't know that I really knew what he might be feeling emotionally," said Steve Seyfried, a classmate. "He was not somebody who would talk about that. He was there to do a job and do it well."

Thomas's classmates also never understood how acutely out of place he felt at Saint John's—even in the shoes he wore. Thomas arrived at school sporting a pair of shiny leather shoes that his friends in Savannah wore. But the white boys at Saint John's, wearing their preppy loafers, ridiculed Thomas's shoes. Thomas ditched the shoes for the type that everyone else at Saint John's had. But when he returned home on the weekends to visit his grandparents, his black friends in Savannah teased him about the shoes and laughed at him for attending "the cemetery," as they called it. Thomas couldn't win. He didn't fit either place, the white world of Saint John's or the black one of Savannah.

In the spring of 1965 he finished his sophomore year at Saint John's. Eleven of his white classmates—half his sophomore class— did not return for their junior year. In the freshman class, Richard Chisolm also elected to drop out, having concluded that he could put up with the school's coldness no more. Chisolm's departure meant that Thomas was now truly alone. No other black student came to Saint John's while he was there.

Thomas viewed surviving his first year and holding his own with his white classmates as a monumental achievement. Although he would face doubts about his abilities again at Yale Law School, he never again believed that he was inferior to any white student. He had proved to himself that he was at least their equal, maybe even their better.

Thomas was only dimly aware of it at the time, but within him a personal philosophy was beginning to take hold that would define the rest of his life. He came to believe that respect, not popularity, was what mattered in life. He stopped caring so much about being popular. Respect was what he wanted. And he would earn that respect by working harder than anyone else, and by posting results—both on and off the field—that would make people take notice.

Three weeks before Thomas returned to Saint John's for his junior year, a California highway patrolman arrested a twenty-one-year-old black man for drunk driving near the Los Angeles community of Watts. What started as a routine traffic incident soon ended in the nation's worst rioting and racial violence in fifty years. At its most heated moments, an estimated ten thousand black youth surged through the streets of Los Angeles. Almost sixteen thousand law-enforcement personnel, including nearly fourteen thousand National Guardsmen, were called in to quell the mêlée. In the end, thirty-four people were killed, six hundred buildings were burned or looted, and property damage totaled $40 million.[4]

The Watts riot of August 1965 was the first of dozens of conflagrations that engulfed major American cities over the next three years. They were, in a sense, the aftershocks of the forces unleashed by the civil rights movement. Americans around the country abruptly woke up and realized that discrimination and racism were not confined to the South. The term *ghetto* suddenly became part of the national vocabulary. A new, more militant generation of black spokespeople burst onto the scene. In California a group of angry black nationalists formed the Black Panther Party. Men and women such as Stokely Carmichael, H. Rap Brown, and Angela Davis raised clenched fists and issued a call to "black power."

In Washington nervous politicians viewed with alarm the surge in "urban unrest." Meanwhile, President Lyndon Johnson steadily committed more and more American troops to the aid of South Vietnam. Both the escalating war and the rising tide of black nationalism sweeping American cities were to have major impacts on the life of Clarence Thomas when he finally reached college.

But at Saint John's, Thomas was far removed from the turmoil of the outside world. When John Lewis and Hosea Williams were being beaten mercilessly by Alabama Guardsmen on the Edmund Pettus Bridge near Selma, Thomas was safely ensconced at Saint John's, both physically and emotionally removed from the raw, racist bigotry. However traumatic the seminary school was for Thomas, it was nothing compared with the physical and emotional abuse that many of his generation were confronting regularly.

What Thomas confronted at Saint John's, however, was the hypocrisy of his Catholic faith. Black children in secular schools held no illusions about what white people thought of them. But Thomas believed, naïvely, that white students studying to be priests would be different. If racial bigotry, however subtle, existed at Saint John's, then the Catholic church's teachings about equality and brotherly love were a sham. For Thomas, the question was basic: How could he belong to a faith that tolerated racial hatred? The question gnawed at him constantly and developed one day into a dramatic break with Catholicism.

Ironically, Father Coleman became one of Thomas's principal links to the civil rights struggle beyond the peaceful grounds of Saint John's. Thomas met regularly with Coleman for his English tutoring sessions, and Coleman told Thomas about some of the momentous changes.

During the summer Thomas earned extra money by working at Camp Villa Marie, the summer camp that still took place on the campus. Every morning the kids and counselors recited the Pledge of Allegiance after morning mass. John Scherer, an upperclassman at Saint John's and a camp counselor, recalled that he and Thomas looked at each another as they recited the words "with liberty and justice for all," mouthing the word *sometimes* at the end.

Thomas's writings from the period reflect his thinking about the deep racial divisions plaguing the country. In 1966 he penned a piece for the Saint John's student newspaper under the headline "It's About Time." He wrote:

Just to keep the records straight, I don't believe that one race is more to blame than the other. . . . There are times when one ethnic group provokes the other until there is a violent eruption of anger and, consequently, more melees, picketing, and rioting.

I think the races would fare better if extremists would crawl back into their holes, and let the people, whom this will really affect, do just a little thinking for themselves, rather than follow the Judas goats of society into the slaughter pens of destruction. True, the intellectuals must start the ball rolling, but ignorance in the intelligentsia is not unheard of.

It's about time for the average American to rise from his easy chair and do what he really and truly believes God demands of him—time to peel off the veil of hate and contempt, and don the cloak of love (black for white and white for black).[5]

When Thomas was in his junior year, Coleman began a program of community outreach, dispatching him and other students to tutor young black children in Sandfly. Sandfly was a sister neighborhood of Pin Point, and some of the families living there were related to Thomas. Because he was black, the children especially looked up to him, some of his classmates recalled, and Thomas enjoyed the experience. It was perhaps the beginning of a lifelong interest in young people, and an aspect of his character that was to distinguish him on the Supreme Court decades later.

Although Coleman pushed Thomas and his fellow seminarians to be brilliant students and committed social servants, he vehemently discouraged another aspect of their development: relationships with girls and anything remotely connected to sex. In keeping with Catholic orthodoxy and the priesthood, Coleman believed sexual temptation was an evil that could lead his young seminarians astray.

Trips home or into town were invariably preceded by admonishments from Coleman to avoid girls and situations that might lead to temptation. At football games in town or at the Saint Patrick's Day parade, Coleman advised seminarians to avert their gaze when the girls' cheerleading squads performed their flips, twists, and splits. He instructed the school librarian to screen magazines such as *National Geographic* for sexually suggestive photographs and censor them with a black Magic Marker.

The sexual taboos produced spirited debates among the seminarians. Was it a sin to look at a girl? At what point did fantasies cross the line from purity to sinfulness?

Now adults, Thomas's classmates laugh about the sexual repressiveness of their alma mater. But many said they were badly damaged by it as young adults, when they left the seminary and began interacting with women their own age. The culture at Saint John's and within the Catholic church imbued them with a deep sense of shame around sex and women. "It was like girls were a different species," said Everson, one of Thomas's classmates. "I just had no idea how to relate to them."

When Thomas returned for his senior year at Saint John's, the eleven members of his junior class were now down to nine after two more defections. Knowing that he was on his way to graduating, he emerged from his shell a bit more. He became the business manager of the school newspaper. He stage-managed a school play. He sang in a singing group. Increasingly, classmates were alerted to his presence by his tremendous belly laugh.

Over dinner one night Thomas told some of his classmates how hard it had been for him at Saint John's. He talked about how no one sat next to him in the dining hall. And how much teasing he had taken over his stupid shoes. In a rare moment of candor, he told them how difficult it had been for him to bounce between his white friends at school and his black ones at home. It was the most that any of his classmates can remember Thomas opening up about himself.

By his senior year, the talk among Thomas and his classmates was also about what to do next. All nine faced the same dilemma: continue with the seminary and the priesthood, or call it quits. Those wishing to continue had a choice among three Catholic college seminaries. One was tiny Conception Seminary College in western Missouri. That year seven graduating seniors decided they would go there. Two classmates bailed out. The freshman class that started with twenty-eight students was now down to seven. Clarence Thomas, a survivor, was among them.

Thomas's graduation from Saint John's in 1967 coincided with a historic change at the Supreme Court. On June 13, President Johnson nominated Solicitor General Thurgood Marshall to become the high court's first black justice, opening a new era in the country's racial politics. Long excluded from power, black Americans were finally getting an opportunity to wield it. The nation's halting effort to make amends for centuries of discrimination became a powerful social force in the years that lay ahead for Clarence Thomas. The Senate confirmed Marshall on August 30, and the crusading civil rights lawyer joined his eight white colleagues on the bench shortly thereafter.

That fall, Thomas took the train to Atlanta and boarded a plane for Kansas City, venturing outside the South for the first time in his life. Founded by Benedictine monks after the Civil War, Conception had been a monastery and abbey before it became a seminary. Benedictine monks served as the faculty and spiritual advisors to the school, leading a contemplative life amid the sweeping farmlands of northwestern Missouri. Like Saint John's, the school was utterly isolated. Kansas City was a two-and-one-half-hour drive to the southwest. Conception Junction, population two hundred, was the closest "town" to the school, followed by Maryville, twenty miles away, with a population of ten thousand. For a black freshman from Savannah, Conception offered the additional extreme of being almost completely white, like more than 99 percent of the people living in the surrounding region. In Thomas's class of sixty-nine freshmen, there were two other black students.

Thomas was already used to being the only black among a class full of whites. But his white classmates, nearly all of whom had attended segregated schools, were not used to having even one black student among them. Tom O'Brien hadn't met a black youth his own age before he met Thomas the first day of school. That day, O'Brien and his friends from Kansas City, who found themselves living on the same hallway with Thomas, tested Thomas with sharp, schoolboy banter. Thomas dished it right back. "Right away, we were pretty much at ease," O'Brien said.

As at Saint John's, life at Conception revolved heavily around the athletic fields. Here again, Thomas won friends and earned respect among his classmates with his overpowering athletic abilities. "I tell you the man back then could throw a football further than anyone I had ever seen," recalled O'Brien. "I swear, we taped off seventy-five yards on a throw once."

At Conception Thomas appreciated for the first time just how hard he had worked at Saint John's. He was better prepared academically than his white classmates, and he found the work at Conception easier than at his alma mater. But Thomas hardly slacked off. "He was ambitious," recalled O'Brien. "We'd hit the books hard for the grade, but he really cared about learning. There would be something extra to read, and he'd be reading it."

O'Brien and Thomas became friends and roomed together midyear. With his parents living in Kansas City, O'Brien invited Thomas home on weekends. The two explored the city together, or headed over to drink beer in Kansas, where the drinking age was only eighteen. They became so close that O'Brien agreed to be the driver for what would have been Thomas's first date. Thomas met the young woman from Kansas City after her choral group sang at Conception. The situation turned sour shortly after Thomas and O'Brien arrived at the young woman's house. Her father opened the door, demanded to know what

Thomas wanted, and ordered him inside. O'Brien heard raised voices coming from the house before Thomas, looking flustered and distressed, emerged alone and raced back into the car. At Conception, the date had told Thomas she was eighteen. Her father informed Thomas that she was only fourteen, a high school freshman, and threw Thomas out on his ear.

As at Saint John's, Thomas's racial antenna was keenly tuned to signals of bigotry. Two overt instances of racism affected Thomas profoundly.

The first occurred at a bar in Overland Park, Kansas, on a night out with O'Brien. They found a booth, and O'Brien ordered a pitcher of beer from the young waiter. They waited for the server to return. Five minutes became ten. Ten became twenty, and still no beer. "What's taking so long?" O'Brien wondered aloud. The manager walked over to the table. "Excuse me, I'm the manager," he said. "We've had complaints. You two have been too loud, and I'm going to have to ask you to leave."

O'Brien was dumbfounded. He and Thomas had hardly raised their voices. He thought the manager was kidding. The manager repeated himself. "You're just wrong," O'Brien told him. "We've hardly said three sentences. It has got to be somebody else. Your guy got it wrong. It's not us."

"I'm going to ask you one more time," the manager shot back, "and then I'm going to call the police. You've been too loud. You need to leave."

O'Brien still wasn't getting it. Thomas leaned over and grabbed his arm. "Man, it ain't the volume," Thomas said.

The manager escorted the two teenagers to the door. Outside, O'Brien looked at his friend in disbelief. Thomas, he saw, had tears in his eyes. "This isn't the first time," Thomas said. "This isn't nothing. It happens all the time."

Like most black southerners, Thomas assumed that racism and bigotry were confined to the South. The barroom incident underscored for Thomas how naïve he was and proved Anderson's wisdom once again.

"Just remember," Anderson had told Thomas before he left for college, "that no matter how many degrees you get and how high you go, the lowest white man in the gutter can call you a nigger."[6]

Thomas's most devastating encounter with racial hatred occurred on April 4, 1968, the day Dr. Martin Luther King, Jr., was shot as he walked across the balcony of the Lorraine Motel in Memphis. Thomas was grief-stricken. In a state of stunned confusion, he mounted a set of stairs behind a white classmate. Someone from below yelled that King had been shot. Without turning around, the student in front of Thomas said, "Good. I hope the S.O.B. dies."

Hours later, O'Brien found Thomas in his room. O'Brien slumped down in a chair. "What's up, Holmes?" he asked Thomas.

"I'm done here," Thomas said. "I'm leaving."

O'Brien stopped short.

Thomas said he thought he wouldn't have to deal with racism in Missouri. "It's not a northern-southern thing," Thomas said. "It's just a thing. It's no better here."

Thomas was through with the seminary, and he was through with the Catholic church. On that night he decided that he could have no part of an institution that produced students who were happy that a black man like King was dead. He also concluded that he could have no part of a religion that preached love and acceptance but practiced bigotry. After years of attending church every Sunday, of rising early to help the priests at Saint Benedict's perform morning services, of studying at Saint John's and committing his life to the Catholic church, Thomas walked away from his faith. At the time, he believed he would never return.

Thomas finished his semester at Conception. But he wasn't coming back. He was going home.

CHAPTER 6

"READY FOR WAR"

Fifteen hundred miles away in Worcester, Massachusetts, Arthur Martin was studying in his dorm room at the College of the Holy Cross when he overheard loud voices in the hallway. "Did you hear?" a white voice shouted. "Martin Luther Coon was assassinated!" Startled, the black sophomore from Newark stepped into the corridor for a drink of water. The white boys abruptly stopped talking and walked away. Martin returned to his room, but he couldn't study. He went and found his friend Orion Douglass from Savannah. It's true, Douglass told him. He's dead.

The two friends sat in stunned silence. As the reality sank in, disbelief turned to rage. "Pissed-off wasn't the word for me," recalled Douglass. "If I didn't have any sense of restraint or discipline, I would have shot somebody that night."

Martin kept asking the same question over and over again. Why King? Why the man who preached love and nonviolence?

"I hated that night like I've never hated in my life," Martin wrote three weeks later. "To me this was the final straw. I was ready to strike back at anything and anybody white. I hated this goddamn country and the entire white race."

Several white students tried to console Martin. He pushed them away. "I told them they couldn't 'feel the hurt or suffer the pain.' No white man could."[1]

On campus, King's assassination gave new urgency to minority recruitment, an issue that had been percolating since the so-called ghetto riots of the previous summer. The ferocity and deadliness of the riots in Newark, Milwaukee, and Detroit caught the nation off guard and turned an uncomfortable spotlight on racism in northern cities. Suddenly, colleges in the north were acutely aware of how few black students were on their campuses. At Holy Cross a handful of black students were pressing the administration to diversify the student body.

A feisty priest of Irish descent named John Brooks, the white chairman of the theology department, was one of the students' strongest allies. Brooks believed Holy Cross could do more for minority students. King's assassination galvanized him to act. Over a period of days he formulated a plan to recruit black students in numbers far exceeding anything the school had done before. It was already April. There was no time to lose.

Without even consulting the college's president, Father Raymond Swords, Brooks and another admissions officer drove to Philadelphia to recruit black high-schoolers to Holy Cross, interviewing about fifty prospective students. They sought a combination of good test scores, strong grades, and solid course work. "I don't think we were looking for Rhodes Scholars," Brooks said. "We were looking for pretty solid citizens who had done okay at school. We figured with a little support here, we could help them along." In the end Brooks offered freshman admission to eighteen black students for the coming fall, pledging tens of thousands of dollars in financial assistance. Afterward, he told President Swords what they had done. "We just offered eighty grand of your money," Brooks recalled telling him.

The 1967/1968 school year was coming to an end, but before the

start of the fall semester the college would make room for one more
black student.

Clarence Thomas said his good-byes at Conception at the end of May
and informed school administrators that he would not return. He
hugged O'Brien and promised to stay in touch. Then he made the long
trip home.

He had told no one at home about his decision. The thought of
breaking the news to his grandfather made him sick to his stomach,
and he prayed that Anderson would understand.

The conversation, such as it was, unfolded worse than Thomas had
imagined. Anderson said almost nothing, listening to his grandson in
silent fury. To him it seemed that his grandson was throwing away an
education and a vocation all at once. Anderson had dreamed of
Thomas's future too long and too hard to accept the news graciously.
In his own anger and dashed expectations, he did the unthinkable: He
kicked Thomas out of his house. "You act like a man," Anderson said,
"you live like one."

Anderson could not have imagined a more devastating punishment.
Home was all Thomas had left. "I was very unhappy," Thomas recalled. "It was like, 'What do you do now?'"

The rift with his grandfather never fully healed. In the end Thomas
and Anderson achieved a form of peace, but it came late in the old
man's life and was never fully settled.

With no job and no money, Thomas moved into his mother's small
apartment in Savannah. After years of being the most focused, disciplined student of any of his contemporaries, Thomas was suddenly
adrift. The future, once so certain, now seemed utterly blank.

King's death was still burning in his heart when Bobby Kennedy,
the first white man at the Justice Department to lift a finger for black
folk, was assassinated in Los Angeles. The world, both at home and
beyond, was falling apart before Thomas's eyes. Everything that had

once seemed so important—getting an education and becoming a priest—suddenly seemed irrelevant. "I stood at the brink of the great abyss of anger, frustration, and animosity," he recalled years later.[2]

Thomas wasn't entirely without prospects that summer. Before he left Conception, he had applied to and been accepted at the University of Missouri in Columbia. But Thomas was through with white schools. He considered some of the South's historically black colleges, but he hadn't applied for admission at any of them.

Walking in Savannah one day in early June, Thomas bumped into Sister Carmine Ryan, his former chemistry teacher at Saint John's. What was he doing? she asked. Thomas told her that he'd left the seminary and sounded vague about his plans.

Sister Carmine was devastated. The nuns and priests who worked in Savannah's black Catholic schools had doted on Thomas. They had worked their whole careers to produce a black student who would rise to the priesthood. The sudden realization that he might fail prompted Sister Carmine to place an urgent call to Robert DeShay, a black Holy Cross student who was home for the summer. "Is there anything you can do?" Sister Carmine asked. "He's too good a person to be left slowly twisting in the wind."

DeShay, Thomas's classmate from Saint Benedict's, was in a position to help. He had worked with the Holy Cross admissions office to recruit black students. He called the office on Thomas's behalf. "I've got a really good candidate for the school," DeShay remembers telling them. The admissions staff made no promises, but they dispatched an application packet to Savannah.

Even after the papers arrived, Thomas remained dubious. There was the practical reality that he was applying so late. Second, he just couldn't get excited about attending another white, Catholic school. In the end he completed the application less out of a desire to attend Holy Cross than out of a sense of obligation to Sister Carmine.

Carmine sent off the application with a letter praising Thomas's academic abilities. In Worcester the admissions office reviewed the ap-

plication. Thomas's grades were good. The fact that he had been training for the priesthood undoubtedly counted in his favor; college officials were well aware of the academic rigors that they entailed. The
school admitted Thomas as a transferring sophomore with a full academic scholarship.

Thomas rode the *Silver Comet* train from Atlanta to New York in the
fall of 1968 with a chicken sandwich packed in a shoe box and one
hundred dollars in his pocket. He and DeShay, a returning junior,
stopped over in New York City to visit DeShay's uncle before boarding a bus for the five-hour ride to Worcester. The trip gave DeShay
plenty of time to fill Thomas in on his new home in the hills of central Massachusetts.

From the mid-nineteenth century well into the twentieth, Worcester built enormous prosperity as a textile town. When Nelson Ambush arrived in the early 1940s, downtown Worcester was alive with
commerce and opportunity. "You couldn't walk a straight line at
lunchtime," Ambush recalled. "There were so many people." One by
one, however, the shirt factories and dressmakers moved south in
search of cheaper labor. By the time Thomas arrived in 1968, Worcester was already in steep decline, following the same slide as other New
England industrial towns such as Lowell and Manchester.

The College of the Holy Cross was built on a hill looking down
the valley to the town. Founded by Jesuits, it had a long tradition of
Catholic education. Latin was a graduation requirement, and philosophy and theology had to account for one third of a student's course
of study. In the years before Thomas arrived, however, administrators
had dropped the Latin requisite, scaled back theology and philosophy
requirements, and begun hiring more and more non-Jesuit faculty.
The obligation of daily mass was also dropped, along with an evening
lights-out policy. Women were allowed in dorm rooms during certain
hours, and the school eliminated the ban on alcohol in rooms.[3]

Although Holy Cross followed Catholic orthodoxy on many social issues, it nonetheless had a moderate reputation. King spoke at the college in 1962, and in 1964, the year after President Kennedy's assassination, President Johnson delivered the commencement address.

The school's lack of diversity, however, had created a difficult atmosphere for its few black students. "You were living in an environment of a tacit or unspoken belief in genetic superiority," recalled Douglass, who had only two other black classmates when he entered Holy Cross in 1964. "If someone of color did excel, it was a freak. It was never mentioned, but it was there. Eventually, it works on you. You kind of adopt it and start subtly believing in it."

But the fall of 1968 was suddenly different on campus. For the first time ever black students didn't feel quite so alone, even though they still accounted for less than 2 percent of the student body. The other colleges in Worcester had also been busy recruiting black students that summer.

Thomas showed up at his assigned dorm room, where he met his new roommate, John Siraco, a white transfer student. On paper the two had little in common. Siraco was studying to be a doctor. Thomas was an English major. Siraco was from the Boston area and had attended segregated, white schools. Thomas didn't have much money and brought no stereo, typewriter, or dorm-room furnishings. He contributed only one decorative touch to the room: a poster of Malcolm X. But both were studious and disciplined. "Clarence was always very considerate, very respectful," said Siraco.

Although Thomas was on an academic scholarship, he still needed money to cover his books and living expenses. Anderson was providing no financial support, even though he could have afforded it. To make ends meet, Thomas took a job waiting tables at Kimball Hall, the college dining room. Elite traditions still survived at Holy Cross, despite recent changes to the curriculum and dorm life. Student waiters served students at mealtimes; maids cleaned their rooms *and* made their beds every morning.

Table waiting dramatically increased the demands on Thomas's
time. He worked five and a half hours a day in the dining hall, in addition to attending three to five hours of classes every day. He also ran
track, which claimed more time in the afternoon. That left evenings
and early morning for studying. He generally stayed in the library until it closed at eleven o'clock, and he was always the first one at the
dining hall for breakfast at seven o'clock. Victor Jackson, another early
riser, said Thomas often logged two hours of studying before breakfast. Years of Anderson's discipline continued to stay with Thomas.

At Holy Cross Thomas was surrounded by black students for the
first time since he'd left Saint Pius School at the age of fifteen. Most
were from northern cities—Philadelphia, Washington, New York, and
Newark—and most were freshmen. Some came on athletic scholarships, others on academic scholarships. All came from strong families
that placed heavy expectations on their shoulders. They were the
cream of their generation: confident, smart, and determined to break
barriers that had truncated their parents' lives.

Fiercely competitive, they sniffed around one another like a pack
of alpha-male dogs, marking territory and testing strengths and weaknesses, freshman Walter Roy recalled. Regional rivalries developed.
Guys from northern cities were constantly dumping on the Savannah
boys, as they were known, teasing them about the South's backward-
ness and ignorance. "We used to call it 'blacker than thou,' " recalled
Leonard Cooper, describing the smugness behind the trash talk.

Thomas moved aggressively to establish his presence. At Kimball
Hall on the second day of school, Thomas sized up one of the
school's star football recruits, a freshman from New York named Eddie Jenkins, while serving him a meal. "You play football?" Thomas
asked. "Let's go. Let's see what you can do." Out on the field, Thomas
took the football and told Jenkins to run. "He threw the ball almost
eighty yards," Jenkins recalled. "He just wanted to challenge me."

Around black friends, Thomas emerged from his protective shell,
cracking jokes and drawing people to him with his explosive laugh.

Over time, Thomas impressed classmates with his academic achievements, but in the early going it was his wit and his athletic abilities that won them over. Thomas quickly developed a close friendship with freshman Gil Hardy, an English major from Philly. Hardy earned the nickname "Three-Nine" because he pulled a 3.9 grade-point average his first semester. Thomas liked smart people and the two friends were constantly arguing, joking, and laughing together. Years later, the friendship became significant for another reason: Hardy introduced Thomas to Anita Hill.

In the fall of 1968 Thomas and other black students organized the college's first black student union. Thomas was elected secretary-treasurer, the number-three post. The preamble to the union's constitution reflected the black nationalist sentiments of the founding members:

"We, the Black students of the College of the Holy Cross, in recognizing the necessity for strengthening a sense of racial identity and group solidarity, being aware of a common cause with other oppressed peoples, and desiring to expose and eradicate social inequities and injustices, do hereby establish the Black Students Union of Holy Cross College."[4]

The BSU lobbied for black professors and administrators, course offerings in black studies, and cultural events that reflected black artistic and creative expression. The notion was simply, Cooper recalled, to create an environment "that reflected that you were there."

For many black students, including Thomas, the BSU became a central part of their campus identity, a place to escape from the cold stares of white classmates. The school gave the union a budget and provided it with a van, which the students drove to off-campus parties with black women, something impossible in Douglass's days. The union also served as a vehicle for political expression, from vehement antiwar protests to militant black nationalism.

It became a hothouse of competition and personal rivalries as well, particularly between Thomas and Ted Wells, a freshman from Washington, D.C., who played football. The two rivals awed their classmates with spirited debates at the BSU on Sunday afternoons. Jenkins recalled them frequently sparring over Booker T. Washington and W. E. B. Du Bois, whose rivalry six decades earlier fractured the black community. Thomas argued for Washington, the guardian of the black underclass, and Wells for Du Bois, the champion of the "talented tenth," the educated and "exceptional" black men who Du Bois believed would lead black people to enlightenment.

Toward the end of Thomas's sophomore year the BSU faced its first internal dispute, over a proposal to establish a dormitory floor reserved for black students. Thomas vehemently opposed the plan, arguing that they had come to Holy Cross to learn with and from white students, not to segregate themselves. How were they going to break down racial barriers by living apart? Thomas demanded. At one point Thomas proposed that at least one black student should sit with white students in the dining hall at every meal, so white students would be forced to deal with black students one-on-one. The proposal went nowhere.

Coming from the South, Thomas was almost genetically wed to principles of integration, a view that was most strongly shared by some of his southern contemporaries. When Orion Douglass learned of the black corridor as a law student in Saint Louis, he was outraged. "We fought for integration," said Douglass. "And when I heard they were going to have the corridor I communicated to Clarence or someone: 'You all are screwing up. People are dying down South for integration, and here you are going into segregation.'"

Yet the sentiment for the black corridor showed how quickly the country had moved in just a few short years. King's call for the "beloved community" in the early sixties was giving way to a more nationalistic orthodoxy. Assimilation was out. Black power was in. "Before a group can enter the open society, it must first close ranks," Stokely Carmichael and Charles Hamilton wrote in their seminal treatment, *Black Power*, in 1967.[5] Proponents of the black corridor wrapped their beliefs in similar rhetoric. "Separatism is the solution for the present race problem in America," BSU member Perry Tin told some two hundred white students at a campus discussion on the proposal.[6]

At Holy Cross, black students were confronting the nitty-gritty of assimilation, and it was leaving a bad taste in their mouth. The indifference of white students was tiring. Black students wanted a place where they weren't constantly feeling as if they had to be on their toes. After King's assassination, there was also a heightened sense about personal safety. No one feared getting shot, but abusive behavior was a lurking concern. Many white students were heavy binge drinkers, and black students felt particularly uncomfortable around the excesses. "You always wondered if it was going to bring out the worst in people," said Stan Grayson. Indeed, a white student rampaged down a dormitory corridor one night screaming racial epithets. The drunken student stormed into the room of a black student and ripped down his black power poster.

When the BSU voted on the black corridor it passed 24 to 1. Thomas was the lone holdout. The next choice for Thomas was whether, once the corridor he fought against was established, he would choose to live on it. He was deeply conflicted. The guys planning to move onto the corridor were all his friends, but he resented their self-segregation. In the end, he decided he would move only if Siraco, his roommate, moved with him. Siraco agreed. Robert DeShay refused to move onto the corridor, out of a sense of loyalty to his white friends, many of whom had supported the push to admit more black students to the college. "I was not going to disaffiliate with those people," DeShay said. DeShay was also becoming annoyed by the rigidity of the student union. "There became this pressure to conform to what the group had decided was appropriate behavior for black students on campus," he said.

Thomas was also beginning to chafe under the pressure to conform. Having lived so much of his life alone, with little emotional support from his family, he felt more comfortable on his own. He was uneasy in groups where he felt he had to give up his own identity. Yet there was also a strong racial component to his thinking. At the seminary Thomas had worked to make the white kids look beyond his skin color.

Next to Richard Wright, no author influenced Thomas's thinking more than Ralph Ellison. He reread *Invisible Man* at Holy Cross, identifying powerfully with the novel's central character, a nameless black southerner locked in a losing battle to assert his individuality free of racial stereotypes. Ellison's opening words captured the experience that Thomas felt as he struggled to find himself in a world filled with negative assumptions and low expectations. "I am an invisible man," Ellison wrote. "No, I am not a spook like those who haunted Edgar Allan Poe; nor am I one of your Hollywood-movie ectoplasms. I am a man of substance, of flesh and bone, fiber and liquids—and I might even be said to possess a mind. I am invisible, understand, simply because people refuse to see me. Like the bodiless heads you see sometimes in circus sideshows, it is as though I have been surrounded by mirrors of hard, distorting glass. When they approach me they see only my surroundings, themselves, or figments of their imagination—indeed, everything and anything except me."[7]

For Thomas, Ellison's words were a fighting challenge. Whatever he did with his life, he vowed that he would never be invisible. He would force people to deal with him as an individual. "Even within this context of we're all black and strong and proud," Thomas said years later, "I wanted to be who I was."

Thomas had never been a slave to fashion, but clothes, ironically, became a central part of his determination to stand out. He wore black leather combat boots, green Army fatigues, a green canvas Army jacket, and a black beret festooned with black power buttons. The wardrobe never varied. He wore different sets of the same clothes day after day, changing into more stylish attire only for evenings out. Sometimes he tucked his fatigues into his combat boots, presenting an almost menacing picture. "When he came looking like that, he looked like he was ready for war," recalled Jenkins.

Thomas's junior year, from 1969 to 1970, was a year of unparalleled turmoil at Holy Cross and college campuses across the nation. Richard Nixon committed thousands more troops to the war in Vietnam, and American casualties were rising steeply. Student protests against the draft and the war reached a fever pitch. More than seven thousand students were arrested during the academic year, and hundreds were expelled for occupying school buildings.[8] It was a year of sit-ins, lock-ins, protests, and marches. It was even a year of death on college campuses. America was at war, both abroad and at home, and college students everywhere were the rear guard of the battle.

For black students in particular, that school year was one of deepening cynicism and escalating confrontation. Many black students came to believe that racism and discrimination were unalterable facts of American life. For Thomas and his classmates, the white hand of discrimination became known simply as "the man." In 1969 the man was everywhere. He was the white kid in the classroom who thought a black student with a brain was some kind of freak. He was the slumlord extracting exorbitant rents from low-income black families. He was the store manager who still believed blacks need not apply. He was the cop on the beat who hassled you because you were the wrong color in the wrong place. And he was the war general who believed black soldiers were cheap fodder for North Vietnamese Communists.

Black students debated actual codes of conduct, a template for how to behave in an all-white environment while retaining a black identity. An undated BSU manifesto from the period listed eleven rules to live by:

1. The Black man must respect the Black woman.
2. The Black man must discipline himself.
3. The Black man must work with his Black brother.
4. The Black man must educate himself.
5. The Black man wants freedom—full and complete.
6. The Black man wants justice—human justice.
7. The Black man wants the . . . opportunity to live and thrive as a human being and the right to perpetuate his race.
8. The Black man wants an immediate cessation of uncivilized civil oppression.
9. The Black man does not want or need the white woman.
10. The Black man must not become involved in the genocidal and inhuman wars of the white man.

And the eleventh command, summing it all up: "THE BLACK MAN WILL TAKE NO SHIT!"[9]

Thomas devoured books by black nationalists—Malcolm X was a favorite—and he adopted much of the rhetoric of the black power movement, complaining bitterly about "the man."

Exactly what Thomas really believed, however, was difficult for classmates to read because he reveled in playing the devil's advocate. "He loved the opportunity to raise an issue and challenge an assumption, and to encourage you to defend a blanket statement," Grayson said. "His debates were not because he had strong convictions on something, it was to see what the basis was for your beliefs." Thomas was never belligerent in his arguments, but there was often an edge to them. "As soon as you thought of him as some guy of ordinary intelligence, or even less, he would then challenge you or come back at you with some big word, and laugh at you," said Jenkins. "Most of the time you didn't believe half the things he was saying."

But some saw definite flashes of Thomas's conservative core, even from underneath his black beret. Thomas was particularly attracted to the half dozen Black Muslims on campus, often joining them for their vegetarian meals, which they ate as a group. Thomas admired their separatist mentality and devotion to self-reliance. The same philosophies drew him to the orthodoxy of the Black Panthers. On the black corridor, Thomas was forever lecturing his classmates about the importance of keeping their noses to the grindstone. And he chided those who blamed poor grades on bigoted professors or disrespectful white classmates. The solution to grades, Thomas believed, was simple: Study harder. Thomas believed that white people ultimately respected only one thing: superior performance. If black students wanted to prove that white prejudices about genetic superiority were dead wrong, they needed to speak the language of accomplishment. Deeds, Thomas believed, spoke louder than words.

In December Thomas played a major role in one of the worst racial crises in Holy Cross history. The crisis developed when members of the mostly white Revolutionary Students Union asked BSU members to help stop General Electric from conducting job interviews on campus. GE was a target because the company manufactured napalm, the flammable, jellylike material that American forces were packaging into bombs and using to defoliate enemy territory in Vietnam. Arthur Martin, the BSU president, said no. Many BSU members, including Thomas, were also opposed to the war, but the war was not the BSU's dominant concern in the fall of 1969. "We had our own Vietnam with civil rights," said Martin.

Catching wind of the planned demonstration, administration officials threatened disciplinary action against any students who violated the college's "open recruitment" policy, guaranteeing students access to prospective employers. On December 10, a Wednesday, about sixty RSU members congregated in front of a room at the Hogan Student Center, where two GE recruiters were waiting. Shouting slogans and linking arms, the RSU members prevented three classmates from entering, forcing GE to cancel the interviews.

Administrators visually identified twelve white and four black student protesters, including Alfred Coleman, a friend of Thomas's from Savannah.

Word of the charges crackled across the campus. All sixteen students faced possible expulsion. That night the BSU called an emergency meeting. Virtually every black student on campus packed into the student-center office. Emotions inside the room were at a fever pitch before the meeting even got under way. Some thought their black classmates had been stupid to participate in the GE demonstration. But everyone was outraged at a glaring inconsistency: One quarter of those charged were black, even though only one tenth of the protesters were black.

Suddenly, someone blurted, "This is a call to arms!" said Malcolm Joseph. The discussion turned to a plan to occupy a campus administration building, by force if necessary, and barricade the building. There was talk of materials necessary to carry out the plan: chains and locks; food and supplies to last days, if necessary.

And then a booming voice from the back wall filled the room. "This doesn't make any sense, man!"

It was Thomas. The room quieted as Thomas outlined a plan to stage a mass withdrawal from school. If the principle at stake was that their brothers were being unjustly thrown out of school, Thomas argued, then the only principled response was for the rest of them to leave as well. Taking over the building was a stupid idea, Thomas argued, because they would all be arrested. "You don't put yourself in a position where you are the issue," Thomas argued, according to classmate Clifford Hardwick.

The more the students talked about the proposal, the more sense it made. They would break no rules this way, and it would create a public-relations disaster for the administration. The school could not allow its entire black student population to leave, they reasoned. "It was a masterstroke," said Martin.

The next day the college disciplinary council debated the fate of the sixteen students. Ted Wells, an aspiring lawyer, demanded amnesty on behalf of the four black students, charging school officials with racism for singling them out. Sometime after midnight Wells and Martin found Father Brooks and told him black students would withdraw if their demands were not met. "It's out of my hands," said Brooks. Wells and Martin returned with the news, telling classmates to call their parents and making plans for a news conference in the morning. Siraco, Thomas's roommate, lent them his typewriter, and he believes that Thomas himself typed their statement.

At three o'clock in the morning, after twelve hours of deliberations, the disciplinary council announced their decision. All sixteen students were expelled. There would be no amnesty.

Shortly before ten o'clock the next morning, black students began making their way to the Hogan Student Center. They carried duffel bags and suitcases. They wore ties and jackets. Malcolm Joseph had a pit in his stomach the size of a bowling ball. He was pre-med. He was going to be a doctor, and here he was about to walk away from it all. Jenkins was risking a pro football career. He remembers wondering if Thomas would show up. Everyone knew that Thomas had big ambitions, too. Would he roll the dice along with everybody else? Jenkins spotted Thomas lugging his suitcase to the news conference.

At ten o'clock Wells read a brief statement to the crowd of students and media. Afterward, the students reached into their wallets, pulled out their Holy Cross identification cards, and threw them on the ground. With raised arms and clenched fists, they walked off the stage and into the December cold. Waiting cars took them off the campus.

The walkout made national news and plunged Holy Cross into crisis. Father Swords, the president, assembled a dozen student and faculty advisors, including Father Brooks. They met most of Saturday and Sunday. Meanwhile, most of the students who left found temporary rooms with students at Clark University across town. Debate inside the administration building was heated. White students on campus formed outside and called for a campuswide strike. They held posters reading SUPPORT YOUR BROTHER. Shortly before six o'clock Sunday night, Brooks recalled, Swords announced that he would grant amnesty to all the students. In a further nod to the concerns of black students, he agreed to publicly condemn "de facto racism" on campus. The crisis was averted.

Though exhausted and emotionally drained, the black students were jubilant at the outcome. It was a defining moment, especially for the architect of the strategy, Clarence Thomas. "Black students became more aware of themselves as black men," Thomas wrote in the school newspaper a week after the walkout. Thomas called the walkout "an action for liberation." Blacks, he said,

> had proven that liberation from the social shackles of racism would be had at any price. Their action demonstrated that nothing is more important than being the black men that they are. . . .
>
> For the black students this Exodus is one more step in the direction of complete liberation from the slavery that whites—whether knowingly or otherwise—persist in foisting upon the black man. Returning to Holy Cross never entered the discussions in general meetings or elsewhere. The black brothers expressed no desire to be a part of an institution that recognized its faults and yet continued to victimize blacks.
>
> Whether or not the blacks went home or not does not really matter. They were willing to forget Holy Cross and make new futures. The blacks acted as men, and that was all that counted. They did not plan to compromise manhood for a "good" education, and didn't.
>
> If a compromising ultimatum had been issued, there was but one answer: forget it![10]

Privately, however, Thomas was far less militant. After the incident, he restricted his activism to off-campus protests and avoided events that might result in his expulsion. "It was too high a price to pay," he said years later. "But I think it was important for them to understand that we thought certain blacks were being treated unfairly."

With racial and political tensions mounting everywhere, Thomas found plenty to campaign against outside Holy Cross. He marched in Worcester to protest discrimination in hiring at a local shoe store. He marched in support of Bobby Seale and the Black Panthers facing murder charges in New Haven. He protested against the Vietnam War. He even joined a contingent of Holy Cross students who marched on the Pentagon in Washington.

Thomas also threw himself into a free-breakfast program in Worcester, started with the help of BSU members and with advice from the Panther Party in Boston. Once or twice a week he rose at five o'clock and caught the bus to a Worcester church. There he whipped up huge skillets of scrambled eggs and pancakes, serving about fifty children every morning.[11]

For Thomas, there was an additional benefit: The secretary/treasurer of the program was Kathy Ambush.

Ambush, a petite woman with delicate features, was the daughter of Nelson Ambush, who moved to Worcester from Boston in the 1940s. Thomas met her at a party in Worcester where she attended a Catholic junior college. A week later, Jenkins recalled, Thomas announced that he was in love.

The Ambush family embraced Thomas, who became a regular at their dining room table on weekends. Thomas played pool and darts with Nelson and took camping trips with the family around the Northeast. It was a sweet time in his life, and the love of the Ambush family helped make up for the heartache he felt about his own family. Thomas was still estranged from his grandfather, and he spoke to his mother only occasionally. His brother, Myers, had joined the Air Force, and the two saw each other rarely. Because of his tight finances, Thomas made no more than one trip home a year while he was at Holy Cross, friends recall.

Now that Thomas had a girlfriend, he became extremely sensitive to the treatment of women on the Holy Cross campus, according to classmates. The school was experimenting with visiting hours for women in the dormitories, and Thomas became one of the toughest enforcers of etiquette on the black corridor. He was constantly moralizing on the need to treat black women with respect when they visited and was particularly hostile toward students who tried to take advantage of women after a night of boozing. He lectured his classmates on controlling their language on the corridor when women were visiting. The directive was a particular challenge for Thomas, as he had one of the foulest mouths on the floor. At one point he penned a poem called "Is You Is or Is You Ain't." The question posed by the verse, Jenkins recalled, was: Is you is or is you ain't a brother, and where you going to do right by the sisters? "Clarence was never part of the promiscuity . . . that was afoot in those days," recalled DeShay. "He always had such high standards about himself, the integrity of his body and what he considered to be the sacredness of a relationship."

Campus demonstrations at Holy Cross and around the country escalated in the spring of 1970. The "Chicago Eight," antiwar protesters arrested during the 1968 Democratic National Convention, were on trial in that city. Bobby Seale and seven Black Panthers were scheduled to go on trial in New Haven, Connecticut, for the alleged murder of an informant, sparking a massive protest scheduled for the May Day weekend. Days before the rally Nixon announced the invasion of Cambodia. On May 4, National Guardsmen killed four student demonstrators at Kent State University.

At Holy Cross officials canceled classes for a week and ultimately canceled final exams for the second time in a year.

Remarkably, the 1969–1970 academic year was stellar for Clarence Thomas. He made dean's list all year and was one of six members of his class admitted to the Purple Key Society, the college's most prestigious student organization. He was also the only black student admitted. "He always told the other guys, 'Do what you want to do with the demonstrations, but don't forget your books,'" Nelson Ambush recalled. "He always did his homework." Thomas's grades were so good that he was named a Fenwick Scholar, a designation given to only a tiny number of students each year. The honor allowed him to design his own course of independent study his senior year.

The crowning achievement of his year, or so it seemed, was when black students elected him president of the BSU. But this prize proved to be short-lived, and its aftermath left Thomas bitterly disappointed.

In a close and intense circle of rivalries on campus, none was more passionate than the one between Thomas and Ted Wells, the spokesman and mediator during the walkout. Fair-skinned and handsome, Wells was smooth where Thomas was blunt. His nickname on campus was "Teddy X," a nod to his militant politics. Whereas Thomas tended toward the shadows of the BSU, Wells always moved in the spotlight.

In school and at the BSU, Thomas competed with Wells over everything. He made no secret of his pleasure at outperforming him. One semester Thomas took a class in economics—Wells's major—just to prove that he could pull better grades, Clifford Hardwick recalled. If Wells was studying late the day before an exam, Thomas loudly informed the rest of the corridor that he, Thomas, was ready for bed. He gloated over the success of the walkout—his idea—and never failed to remind his classmates what might have befallen them if they had padlocked the administration building, which, according to Hardwick, was Wells's idea. At BSU meetings Thomas was constantly picking apart Wells's proposals or statements. To some it seemed as though Thomas simply took the opposing position just to provoke an argument with Wells. "It was a game to them," Hardwick said.

There was a subtle racial edge to the rivalry, especially for Thomas, who regarded Wells as part of the elite, light-skinned class of black Americans that had snubbed him in Savannah. In fact, Wells came from a working-class background, raised by a mother who was a postal clerk for the Navy.

In Thomas's junior year, the BSU met to elect a new president. It was widely assumed that Wells, the BSU's vice chairman, would replace Martin. But on the evening of the election, Clarence Thomas was the top vote-getter.

The victory was the capstone to his year. His classmates had chosen the poor kid from Pin Point over the pretty boy from Washington, D.C. The vote seemed to validate Thomas's spin on their rivalry.

Almost immediately, however, some BSU members began to have second thoughts. Some members had voted for Thomas as a way to show up Wells, who some believed arrogant. Doubts set in about Thomas, who had never been as closely involved in the nitty-gritty of the BSU's administration as Wells had been. As Jenkins recalled, Thomas was a "bomb thrower," not a serious tactician. Thomas's image was also an issue. His uniform wardrobe reflected their radicalism but not their elitism.

A contingent of BSU members campaigned to reconsider the vote. At the next meeting, Hardwick made a motion to do so. Thomas, who could have objected, let it go forward, and Wells became president of the BSU.

Thomas told others he wasn't peeved. But those close to him saw a change. "That hurt him," said Victor Jackson, a Savannah classmate. "I think he lost something there." After the election, Thomas withdrew from the BSU and became more critical of its decisions. He turned his attention to his schoolwork and focused on applying to law school. "After the election, I think he decided, 'Now I need to look out for myself,'" Jackson said. "I think he felt determined that he was going to show people that he was superior, that he could succeed."

The defeat also deepened Thomas's dislike of group-imposed limits. Like his friend Robert DeShay, he began to rebel against the group psychology that mandated conformity. "Clarence is one who always resisted the notion that someone else was speaking for him, that because he was five foot ten he must think this way, or because his eyes were brown he must do this in this situation," said Stan Grayson. "In a period where there were a bunch of people trying to be the voice of black America, I think that was troubling to a lot of people, but certainly for him."

Thomas's epiphany came during a protest rally in Boston to support the Black Panthers and other "political prisoners" in the spring of 1970. In the din of angry shouting, Thomas heard a voice inside his head question the utility of group protests. Why was he on the street, instead of getting an education? Why was he angry, instead of heeding God's call to love? Suddenly, he believed racial anger and bitterness were leading him away from everything he'd worked for.

"It was intoxicating to act upon one's rage, to wear it on one's shoulder, to be defined by it," Thomas recalled years later. "Yet, ultimately, it was destructive, and I knew it."[12]

Deep down, Thomas was also never completely comfortable protesting, marching, and expressing himself so confrontationally against the status quo. The discomfort with direct confrontation was too deeply ingrained from his years living in Georgia under segregation, where conflict avoidance was the norm, and studying for the Catholic priesthood, which taught pacifism rather than activism. Militancy suited Thomas as poorly as his ill-fitting combat fatigues, and in the end he shed them both.

At Holy Cross Thomas vowed to do what he believed was the most radical statement he could make as a young black man: finish college and make something of his life.

The only question was, what?

Ironically, Thomas's decision to become a lawyer was essentially made by default. Growing up in Savannah, Thomas had heard of only two worthy professions from his grandfather: the priesthood and the law. For Anderson, both were deeply rooted in the realities of Savannah in the 1950s and 1960s. Savannah had almost no black lawyers, just as it had no black priests. Under segregation, none of the all-white state law schools accepted black students, and none of the state's historically black schools offered the law as a graduate degree. In the 1950s and early 1960s Savannah had two prominent black attorneys, and both were consumed by litigating civil rights cases, such as the desegregation of the school system. Ordinary black citizens such as Anderson depended on a handful of white attorneys who were willing to represent them. "We didn't have a good lawyer here," said Sam Williams, Anderson's business partner and friend. "None of the whites would accept the cases, or it was very rare." A young Savannah attorney named Bobby Hill also inspired Thomas toward the law. In 1968, the summer Thomas returned from Conception, Hill was elected to the Georgia General Assembly, the city's first black representative since Reconstruction. Hill's victory was widely heralded in Savannah's black community, and Hill himself became a local icon.

At Holy Cross Thomas was surrounded by classmates who were heading toward legal careers. Martin, who graduated the year before Thomas, went to Georgetown Law School. On the heels of the civil rights victories of the 1960s, many young black men and women regarded the law as the way they could make a difference. "The law was prominent in our lives," recalled Grayson, who also went from Holy Cross to law school. "There were all kinds of issues every day about the rights of people of color." The law also seemed to offer the opportunity to make good money. "To the extent that we were all striving to have a better life for ourselves than our parents had, which was certainly their desire, the law kind of assured everybody a minimum standard of living," said Grayson.

Shortly before graduation, Thomas told Nelson Ambush that he wanted to earn enough money one day to afford a Rolls-Royce. "He had it in his mind that he wanted to do something with his life that really counted," Ambush said. "He was going places, and he was going to amount to something."

With the LSAT's behind him and his law-school applications in the mail, Thomas began to loosen up during his senior year. He continued to log the same hours at the library during the week, but by Friday he was ready to kick back with a bottle of Boone's Farm wine or a few Schlitz beers. He partied at some of the neighboring schools in western Massachusetts. Years later Thomas admitted to trying pot. Increasingly, he spent time with Kathy Ambush, and no one was surprised when the two became engaged. They scheduled the wedding for June 5, 1971, the day after Thomas was to graduate.

More good news came in the mail. Thomas was accepted to three of the nation's top law schools: Yale, Harvard, and the University of Pennsylvania. Because of his grades, the Holy Cross faculty also recommended him for membership in Alpha Sigma Nu, an elite honorary society of Jesuit scholars. Thomas was one of twenty in his class of about six hundred to be selected.

Thomas took a black-literature seminar with classmates Eddie Jenkins and Gil Hardy that, according to Jenkins, became allegoric of their experience at Holy Cross. Two poems in particular represented the often competing extremes of their existence: "If We Must Die" by Claude McKay and "We Wear the Mask" by Paul Laurence Dunbar.

McKay's poem was a call to arms, the words exploding with fury in every verse:

> If we must die, let it not be like hogs
> Hunted and penned in an inglorious spot, ...
> If we must die—O let us nobly die, ...
> Pressed to the wall, dying, but fighting back![13]

In sharp contrast, Dunbar's poem captured the rage and anger that blacks suppressed around whites. "We wear the mask that grins and lies,/It hides our cheeks and shades our eyes."[14]

In a sense, Thomas and his classmates saw McKay's poem as the message of the black power movement, Jenkins recalled. "One was a clear call for revolution: that we would not allow people to just take our humanity away," he said. "Before they'd do, we'd die first." They saw Dunbar's poem as a survival tactic. "Dunbar talks about how we must go through society and disguise the real guile that makes us an-

gry and bitter," said Jenkins. "If you want to survive in white America, you have to develop that face."

By the end of the course, it was clear to Jenkins which verse spoke
loudest and clearest to Clarence Thomas. "The Dunbar poem was
what every African American at Holy Cross had to choose," he said.
"If you wanted to become a lawyer, doctor, or chief executive, how
could you choose Claude McKay? It was death."

On June 4, 1971, Thomas graduated from Holy Cross. At the age
of twenty-two, he had scaled a height that no other member of his
family had achieved. He graduated cum laude, indicating a grade-
point average of better than 3.5. Michael Harrington, the author of
How the Other Half Lives, a foundational work in the antipoverty programs of the 1960s, delivered the commencement address. The following day Thomas and Kathy Ambush were married at a Worcester
church. Gil Hardy was Thomas's best man.

At both events, Thomas's family was represented by his mother and
his grandmother. There was one notable absence. Myers Anderson,
still angry at his grandson for leaving the seminary, declined to make
the trip north for his graduation or wedding. The man who had so
much to do with Thomas's making it to Holy Cross, the man who had
sweated, prayed, saved, cajoled, and threatened for the sake of his
grandson's education was too proud and too stubborn to see him actually succeed.

But Thomas had other plans. He had a new wife, and he was heading to Yale Law School. As he told the alumni newsletter before graduation, he was eager to return home to Savannah with a Yale law
degree and "begin helping some people."[15]

"That's why I finished college," he said.

An undated photograph of Myers Anderson, the grandfather who raised Thomas, reveals the stern, uncompromising face he displayed at home. He was a pillar in Savannah's black community, and his acts of charity became local legend.
(*Courtesy of Leola Williams*)

Leola Williams (*right*), Thomas's mother, was raised by relatives in Pin Point, Georgia, outside Savannah after Anderson abandoned her. Williams married Thomas's father, M. C. Thomas, as a teenager and had three children before divorcing him in 1951.
(*Jonas Jordan/Associated Press*)

Two years Thomas's senior, sister Emma Mae Martin stands in front of her home in Pin Point. Unlike her two brothers, Martin was raised by relatives in Pin Point after her parents divorced.
(*Stephen Morton/Special for the* Atlanta Journal-Constitution)

Leola Williams, daughter Emma Mae, and granddaughter Leola Farmer stand outside the house that Myers Anderson built in the 1950s in Savannah. Clarence and brother Myers Thomas shared a bedroom there during the ten years they lived with their grandparents. (*Jenni Girtman*/Atlanta Journal-Constitution)

After attending an all-black Catholic parochial school in Savannah, Thomas gained admission to an elite private seminary outside Savannah for his final three years of high school. This senior-year photo shows him as the business editor of the school newspaper. (*Catholic Diocese of Savannah*)

The first to integrate the seminary, Thomas was the only black student in his class. Extremely hardworking, he was considered shy and reserved by classmates. (*Catholic Diocese of Savannah*)

A gifted athlete, Thomas made friends and impressed classmates with his performances on the basketball court, baseball diamond, and football field. He remains passionate about football and basketball. (*Catholic Diocese of Savannah*)

Thomas with his Washington mentor, J. A. Parker, a noted conservative, on the day he was sworn in as chairman of the Equal Employment Opportunity Commission in 1982. Thomas was chairman of the EEOC until 1990, the longest-serving chairman in the agency's history. (*Equal Employment Opportunity Commission*)

Thomas was a lightning rod for controversy at the EEOC but transformed the agency into the modern, professional bureaucracy that it is today. A close-cropped goatee in this photo from the mid-1980s is the only reminder of the radical politics Thomas espoused in college. (*UPI*)

Thomas testified before House and Senate committees more than fifty times while at the EEOC. Difficult encounters with Congress like this one in the mid-1980s set the stage for his contentious nomination to the Supreme Court in 1991. (*Equal Employment Opportunity Commission*)

President George Bush secretly flew Thomas to Kennebunkport, Maine, on July 1, 1991, to offer him the job of succeeding Thurgood Marshall on the Supreme Court. Here the two men met privately in Bush's bedroom several hours before Bush officially nominated Thomas. (*Associated Press*)

Thomas defends himself against Anita Hill's allegations of sexual harassment on October 12, 1991, the second day of the reopened hearings into his nomination. Behind him are his wife, Virginia; son, Jamal; mother, Leola; and sister, Emma Mae.
(*Rick McKay/Cox Newspapers*)

Thomas poses with his brother, Myers, and nephew, Derek, the day he took the constitutional oath of office at the White House. Myers Thomas died of a heart attack in 2000.
(*Abraham Famble*)

Despite differences in ideology, Thomas works well with all his colleagues and is especially close to Justices Antonin Scalia and Ruth Bader Ginsburg, as both were his colleagues at the U.S. Court of Appeals for the District of Columbia. Front row, from left, are Associate Justices Scalia,. John Paul Stevens, Chief Justice William H. Rehnquist, Associate Justices Sandra Day O'Connor and Anthony Kennedy. Back row, from left, are Associate Justices Ginsburg, David Souter, Thomas, and Stephen Breyer. (*J. Scott Applewhite/Associated Press*)

Off the bench, Thomas regularly meets with schoolchildren and has developed one-on-one mentoring relationships with dozens of young people over the years. Mario Scott, the young boy behind Thomas, came from a broken home; Thomas arranged for him to attend a private military academy in Virginia. (*Rick McKay/Cox Newspapers*)

Thomas becomes emotional during a speech in May 2001 while discussing his adoption of his grandnephew. The boy's father, Thomas's nephew, dealt crack cocaine in Savannah and was sentenced to a thirty-year sentence after his third drug conviction in 1999. (*Jenni Girtman*/Atlanta Journal-Constitution)

After his speech in Savannah, Thomas visited with two childhood friends, Isaac Martin (*left*) and Abraham Famble (*right*). Martin's son, Isaac, was also sent to jail in the sting that netted Thomas's nephew. (*Jenni Girtman*/Atlanta Journal-Constitution)

Virginia Lamp married Thomas in 1987. Raised in Omaha, Nebraska, she met Thomas at a conference in New York. Friends credit Virginia with pulling her husband through the worst moments of his Supreme Court confirmation. The couple sent out this photo as part of their holiday greeting card in 2003.
(*Courtesy of Virginia L. Thomas*)

Thomas is an avid NASCAR fan and a lifelong auto nut. He learned auto mechanics from his grandfather and owns a Corvette ZR-1. He also owns a forty-foot luxury mobile home that he drives on family vacations. In 1999, Thomas served as grand marshal of the Daytona 500.
(*Courtesy of Virginia L. Thomas*)

Thomas adopted Mark Martin, his grandnephew, in 1997. He leaves the Supreme Court midafternoon to pick Mark up from the bus stop and often challenges him with extra homework assignments. Starting a family over again became an important part of Thomas's healing after his humiliating Supreme Court nomination. Thomas also has a grown son by his first wife. (*Courtesy of Virginia L. Thomas*).

CHAPTER 7

IVY LEAGUE FAILURE

The stench of rotting garbage hung thickly in the stone corridors of Yale Law School when Clarence Thomas arrived in June 1971. The preceding spring the university's janitors, food-service workers, and support staff had struck for higher wages, ending their walk-off only weeks before the start of fall semester. Piles of uncollected garbage, still marinating in the summer heat, were only just being hauled away as students settled into their dorms. Clarence and Kathy took off-campus housing in a small New Haven apartment. Kathy got a job as a bank teller, while Thomas earned money interviewing walk-in clients at a New Haven legal-aid office.

About to begin classes at the most prestigious law school in the country, Thomas should have felt nothing but the thrill of anticipation. Instead, he felt hopeless and confused. How had the world gotten so far out of whack? *The New York Times* had just begun publishing the Pentagon Papers, disclosing years of lies and deception in the American-led war in Vietnam. Thomas picked up a copy of Marvin Gaye's new album, *What's Going On*, and, sitting in his tiny one-room apartment, played the title cut over and over again. The notes of Gaye's unforgettable voice touched something in Thomas, and his words of war, death, "picket lines, and picket signs" seemed to capture the turmoil of his generation. "War is not the answer," Gaye sang.

"For only love can conquer hate." Having concluded that the politics of anger were ultimately destructive, Thomas was left with an inner void. What was the solution to the chaos? Was it something as simple as love, as Gaye said? In a sense, Thomas came to Yale looking for answers to these questions, as much as for an education in law.

At the time Thomas could not possibly have foreseen the ways in which Yale would impact his life. The contacts he made there helped catapult him to the steps of the Supreme Court, then nearly tripped him as he reached the threshold. The school also figured prominently in Thomas's political journey, taking him farther from campus liberalism and closer to conservatism. Yale's elitism reawakened Thomas's deep resentments over class and racial distinctions. He made career decisions that defied expectations of what a black lawyer should do.

And for the first time in his young adult life, Thomas tasted failure, or so he thought, when he lost out on the job of his dreams after graduation. The experience, aside from nearly devastating him, figured prominently in Thomas's hostility toward affirmative action. Today, Thomas's opposition to affirmative action, although he benefited from it at Yale, stokes much of the anger toward him among many Americans, black and white.

Thomas's decision to attend Yale was partly based on his desperate financial situation. The school offered him the best aid package, and he was in no position to turn it down. From a distance, Yale also appeared far more hospitable toward black students than other Ivy League schools. Yale in the 1960s and 1970s was a bastion of East Coast liberalism. Two men burnished the school's liberal image. The Reverend William Sloan Coffin, Jr., Yale's chaplain, was a veteran of the freedom rides of the early 1960s and a passionate civil rights activist. He was also one of the most outspoken critics of the Vietnam War at any American university, Ivy League or otherwise.

Kingman Brewster, Yale's president, locked horns with Nixon over the Vietnam War and electrified the nation when he declared his skepticism that "black revolutionaries" could receive a fair trial in America, sentiments he uttered on the eve of the Black Panther trials of 1970. Front-page news around the nation, the comments drew angry rebuttals from conservative Yale alumni, many of whom cut off donations in protest. Alumni hostility toward him became so intense that Brewster later joked that he could solve Yale's chronic financial troubles by auctioning off his resignation. Yet Brewster's liberal rhetoric signaled to millions of black Americans that he understood something about race in America.

In reality, however, Yale's track record with black students was little better than any other northern university. At the law school, black students comprised only 1 percent of the student body before the 1967–1968 academic year, when black enrollment increased to just 2.6 percent.[1] The law school did not hire its first black administrator until 1969. The Yale Corporation, the powerful governing body of the university, did not elect its first black member until 1970, 269 years after the school's founding. That member, a graduate of the law school, was a district court judge from Philadelphia named A. Leon Higginbotham, Jr., who much later became one of Clarence Thomas's most virulent critics.

The same forces that drew Thomas to Holy Cross were at work at Yale Law School. In 1969, the year after King's assassination, the school more than doubled the number of incoming black students, to about fifteen. The number was doubled again after black students on campus denounced the recruitment measures and demanded greater representation at the law school. The incoming first-year class in 1969 had more than thirty black students. By 1970, however, more than half of those students had flunked out.

The failure of so many students plunged the law school into a crisis that was left to the school's new dean to solve. Abe Goldstein was sympathetic to the plight of black Americans trying to crack the barriers of a graduate education, yet deeply mindful of Yale's traditions. He wanted more black students, but he would not compromise the school's standards to get them. Working with an outside consultant, Goldstein and his staff developed an admissions formula for minority applicants that relied heavily on LSAT scores and grades. Applicants were admitted if the formula projected that they could achieve the equivalent of a C-plus grade-point average. White and black applicants were evaluated separately, and a "modest" grade and test-score differential was used to make the final cuts, Goldstein said.

Ralph Winter, a professor who served on the minority application committee, said black students were evaluated the same way the school would consider the children of important alumni, commonly known as legacies. "The test would be the same," Winter said. "Did they stand a chance of success? If so, they got in."

"It was built on the naïve assumption that this was a temporary problem," Goldstein recalled. "We were quite unsophisticated."

Clarence Thomas and his classmates were admitted to Yale under the new formula. It was the beginning of systematic affirmative action at Yale Law School.

Thomas's freshman class included 160 students. Twelve, including Thomas, were black. Regardless of race, they were among the best law students in the country. While Yale and Harvard argued over which school was the best, there was no argument over which was the most exclusive. Yale Law's student body of just over five hundred students was only slightly larger than the typical first-year class at Harvard Law. The school capitalized on its small size by cultivating an elite image. At Yale there were no letter grades or class rankings. Such distinctions were thought meaningless for students who had proved their abilities simply by being admitted. It looked for students who showed promise not just as lawyers but also as citizens. Students went to Harvard to become lawyers, Yale alumni liked to say. They went to Yale to become leaders. "The assumption was that everyone would be flipping back and forth between large law firms and government," recalled David Jones, one of Thomas's classmates.

The ethos of the school appealed to students from all over the country, black and white, rich and poor. It was what drew a poor white kid named Bill Clinton from Hope, Arkansas, to Yale in 1970. And it was partly what drew a poor black kid from Pin Point, Georgia, there in 1971.

The school boasted some of the top legal minds in the country. Louis Pollak, the outgoing dean when Thomas arrived, helped build the plaintiff's case in the historic *Brown* decision. Alexander Bickel defended *The New York Times* in the Pentagon Papers case. The bible of civil procedure, *Moore's Federal Practice*, was authored by James William Moore of the Yale faculty. Burke Marshall, a former Justice Department lawyer under Presidents Kennedy and Johnson, defended the constitutionality of the 1964 Civil Rights Act. Eugene Rostow, a veteran of the State Department under President Johnson, also taught at the school. Thomas Emerson, known among students as "Tommy the Commie" because of his staunch liberalism, successfully argued a Connecticut birth control case that became the foundation of the Supreme Court's landmark *Roe* v. *Wade* decision legalizing abortion. A young conservative professor named Robert Bork was also making a name for himself in First Amendment law.

Whatever Thomas's own expectations about Yale, he quickly learned that the school had expectations of its own. For John Doggett, a black lawyer who graduated ahead of Thomas, the point was driven home one day when a classmate complained about the absence of class rankings at Yale. "You are the cream of the crop of law schools in the world," Doggett recalled the professor's response. "You do not separate cream from cream. It is your fate as a Yale Law School student to become one of the leaders in the legal profession. It will happen, not because of you personally, but because you are here. That is what happens to Yale Law School students."

Thomas nearly collapsed under the weight of Yale's expectations his first year. In some classes he hyperventilated at the thought of being called on. Yale's Socratic method of teaching law, so dramatically portrayed in the movie *The Paper Chase*, reminded him of Coleman's Latin classes at Saint John's. All his old demons about genetic inferiority suddenly resurfaced around so many white classmates. Used to being near the top of his class at Holy Cross, he suddenly confronted the perception that he was merely average, at least compared with the brilliant minds around him. During spring semester he took a property law class from one of the toughest teachers at the school, Quintin Johnstone, and received a "low pass," just above a "fail." Maybe he wasn't cut out for Yale, he began to think. Contemplating dropping out, he told Johnstone of his frustrations. Johnstone talked him out of it.

Since he lived off-campus, Thomas interacted with his classmates mostly in the classroom or the dining hall. White and black students studied together at Yale, but they still tended to socialize and eat separately, almost unconsciously. In the dining hall, "somebody black would go sit down, then other black people would sit down, and you reach a critical mass and no white people sat down," recalled Wendy Samuel, one of Thomas's black classmates. "I don't think anybody felt we were being pushed to do that. It was more the other way around. It's what we did to feel comfortable."

Thomas mixed freely with both white and black classmates but behaved very differently around each set of friends. To white classmates, Thomas seemed shy and reserved, the guy who habitually slouched in the back of the classroom. To his black classmates, however, he was gregarious and outgoing, constantly making himself the center of attention. He overwhelmed his classmates with his trademark laugh and endless jokes. "Clarence was an incredibly funny guy, very witty," said classmate Dan Johnson. "He had a bag of jokes you wouldn't believe. The guy was constantly coming up with new stuff. He kept me in stitches." As he had at Holy Cross, Thomas argued relentlessly, sometimes seriously, sometimes mockingly. "There was nothing politically correct about Clarence," said Johnson. Around black classmates Thomas could relax and be himself. Around fellow whites, he was as unnoticed as Ralph Ellison's invisible man.

Race cut deeply for Thomas at Yale. He clung fiercely to his identity as an "authentic" black man, nearly always sitting at the black table in the dining hall. Indeed, Thomas saw sharp differences between himself and black classmates, some of whom were light-skinned and from privileged backgrounds. To some, it seemed, Thomas adopted the "blacker than thou" attitude exhibited by the big-city boys at Holy Cross. In political discussions he was constantly poking fun at the black bourgeois attitudes of classmates. The humor had a certain charm to it, classmate David Jones recalled, but also a certain edge. "Clarence was very irritated, and he covered it with humor, as to what he saw as his climb up and the presumptions and attitudes of the black bourgeoisie who had come up on the basis of the civil rights movement," said Jones. "He felt that the civil rights movement and the sort of 'oomph' that had come from it hadn't really benefited people like him. This class issue really came through for Clarence."

Thomas identified most closely with classmates Frank Washington and Harry Singleton, who both came from poor backgrounds and were the first members in their families to go to college. "We all felt that, not having the benefits of some of our classmates, we had been able to start the race from a hundred yards back from the starting line but finish up in the same place," said Washington. Thomas and Washington had long discussions about the discrimination dark-skinned blacks faced within the black community. Thomas told Washington of the teasing and prejudice he faced in Savannah. And he related some of the stories about how even his grandfather was looked down on by the wealthier, fairer-skinned black families.

Thomas said he simply felt out of place at Yale.

"I didn't belong there. I didn't fit in. I was unhappy," Thomas said, reflecting on his experiences years later. "I was twenty-something years old. I was married. . . . I was working. I had no money. I had no clothes. How was I supposed to feel? I'm working every hour I can at New Haven Legal Assistance. I live in a little, dumpy one-room apartment. It wasn't dumpy, but it was one-room. I see all sorts of injustices in society. How are you supposed to feel?"

As he had at Holy Cross, Thomas wore his convictions on his sleeve. Ditching the Army fatigues, combat boots, and black beret he wore at Holy Cross, he instead dressed in bib overalls and a wool knit cap, reminding Professor Johnstone of a "southern farmer." The clothes were a deliberate statement about Thomas's roots and political sympathies, said Jones. "He was wearing the overalls not because he had to, but to make a point: 'I made this all by myself.'"

Thomas also noticed the difference between himself and many of his classmates in his wallet. Yale was paying his school bills, but no one was covering his living expenses. He needed the job at New Haven Legal Assistance to make ends meet. He owed the job to a white Republican judge he'd never met. Before Thomas arrived at Yale, Father Brooks at Holy Cross called Judge Angelo G. Santaniello, a Holy Cross alum, and asked him to help Thomas and his wife find work in New Haven. Santaniello contacted his friend Fred Danforth, who ran the New Haven Legal Assistance office.

The New Haven Legal Assistance Association, or LAA, maintained several neighborhood offices in poor New Haven communities. The jobs were highly coveted among Yale law students for the practical experience in plaintiff law. Those seeking the jobs, however, were overwhelmingly white, while LAA's clients were mostly black. "It was very hard to get black lawyers out of the law school because they would be offered all these fancy jobs," said Danforth. Thomas was the answer to a prayer.

Danforth assigned Thomas to LAA's Grand Avenue office, near a large public-housing project. Thomas became the point of contact for people walking in the door. He interviewed clients, figured out what their legal issues were, and recommended priority cases for action. "He was very good at that," said Frank Cochran, the other attorney in the office. "He was a very quick study on people coming into the office."

Danforth and Cochran were so taken with Thomas that they promised him a job after graduation, offering a salary of about fourteen thousand dollars a year. Thomas turned them down. They offered more money. Thomas turned them down again. Cochran said he just assumed Thomas was interested in the work, based on his obvious ability to relate to "folks on the street."

"But he was not," said Cochran. "He just made that point politely, but very clear."

At the time Thomas still had his sights set on Savannah. A legal-assistance job in New Haven wasn't part of the game plan.

Thomas returned for his second year at Yale in the fall of 1972 with some exciting news: He was going to be a father. Kathy was expecting a baby in February. Thomas also delighted in the company of Gil Hardy, who arrived at Yale in 1972 as a first-year student.

Academically, Thomas began to find a groove. He took another property-law class from Johnstone. He was interested in property law, but he also viewed Johnstone as a challenge. He wanted to prove he could do better than his low pass from the year before. Thomas didn't avoid weaknesses; he tackled them head-on. He didn't earn the top grade from Johnstone that year, he said, but he did the next year when he took another class from his favorite professor.

Because he was married, Thomas didn't participate as much in the limited social scene on campus. He played pinball with David Jones. He joined an intramural touch-football team, and did his part for the old Blue when a team representing the *Yale Law Journal* crushed one representing the *Harvard Law Review*. Between classes and his family, there was time for little else. The law school had its own film society, and one of the big attractions on campus was Sunday-night movies in the law-school auditorium. "Blue-movie night" was an especially big draw. In the early 1970s pornographic movies were just beginning to hit the mainstream. *Deep Throat,* starring Linda Lovelace, and *Behind the Green Door* with Marilyn Chambers were both released in 1972 and became cult classics. Hundreds of law students, male and female, crowded into the law-school auditorium for the blue-movie nights. Thomas went, too, usually chewing on a big cigar, recalled Dirk Schenkkan. "We were just all reveling in the sexual freedom of the time," said Schenkkan. "It was the era of liberation."

The laissez-faire attitude toward pornography was indicative of the general climate of liberalism that pervaded the law school. It was the era of free speech and individual rights, of tearing down social barriers and of unlimited indulgence. The law school in particular was unabashedly liberal. "I'm sure there were some conservatives around," recalled Schenkkan, "but I don't think they announced themselves."

Liberalism became the dominant filter through which students saw world events. Nixon was regarded as a reactionary law-and-order president, and the Vietnam War as immoral and unjust. By contrast, the Supreme Court was viewed as the most progressive branch of the federal government. Liberal decisions were expected from the Court. Even the historic 1973 *Roe* v. *Wade* decision legalizing abortion wasn't viewed as particularly noteworthy at the time, according to some of Thomas's classmates.

Roe later surfaced as a major issue in Thomas's confirmation when he told the Senate Judiciary Committee that he hadn't really given the decision much thought. "When Clarence talks about not having really talked about *Roe* v. *Wade* or analyzed it, he's actually right," said classmate Tim Terrell, a law professor at Emory University. "That was not a big conversational piece because at that point the Supreme Court was on a roll continuously of inventing and expanding individual rights under the Bill of Rights."

The school's liberal character came through most strongly, perhaps, in how the law was exalted as a tool for social change. "Pretty much everybody thought that the law could be used to remake society in a more just and progressive way," said Don Elliott, a classmate of Thomas's who later taught at the school. In the Yale worldview, the federal government held ultimate power to solve society's ills and deliver justice. "States were irrelevant," said Elliott.

Students learned that the federal government's "plenary powers" superceded state or local laws. The federal government derived this authority, Yale Law students were taught, from the Supreme Court's reading of the commerce clause in the United States Constitution. The clause gives Congress the power "to regulate Commerce with foreign Nations, and among the several states, and with Indian Tribes." According to Elliott, "We were basically taught that the meaning of the commerce clause was that there was no stopping point in terms of federal power."

Thomas bucked the conventional wisdom, arguing that such an expansive reading of the commerce clause was inconsistent with the powers and authority granted by the Constitution to the states. He argued the point strenuously in class, Elliott recalled. "I can remember Clarence making the argument that . . . even if there was no linguistic stopping point under the commerce clause, that the commerce clause had to be read together as part of a totality, with the structure of the Constitution and the role that states were given."

Two things were striking about Thomas's argument. It was totally contrary to the conventional thinking of the time, particularly at Yale, and it ultimately became the position he adopted years later on the Supreme Court. It was, perhaps, Thomas's first step toward conservatism.

Yet no one, least of all Thomas himself, would have characterized him as conservative at that time. Still deeply wedded to the liberal positions of his Holy Cross days, Thomas despised Nixon and remained deeply opposed to the Vietnam War. And when someone gave him a new book by a conservative black economist named Thomas Sowell, Thomas threw it in the trash. To Thomas, Sowell was a sellout, what

at the time he called a "half-stepper," a black man of means who ac-
commodated with white power. Fewer than three years later Thomas
came to view Sowell in a radically different light.

Ever so subtly, however, Thomas's class sensitivities were beginning
to steer him away from liberal politics. Yale's overbearing liberalism
also stirred his natural contrariness, and he grew increasingly dubious
about the leftist rhetoric he heard from classmates. Frank Washington
said he and Thomas came to view liberalism as "dilettantish affecta-
tion for people who had the money and could afford that."

"We viewed the world much more pragmatically," said Washington.
"It wasn't that we had made a decision that liberalism was bad or
good. A lot of these people, they could afford to take a European va-
cation, whereas people like us knew we had to get a job."

On February 15, 1973, the day after Valentine's Day, Kathy gave birth
to a son. In a nod to Thomas's black nationalism, they named the boy
Jamal Adeen Thomas.

Fatherhood and the responsibility of raising a family suddenly
made gainful employment paramount. Thomas, along with many of
his black classmates, was considering a career in civil rights law and
lined up a summer job working for his Savannah idol, Bobby Hill.

Seven years Thomas's senior, Hill graduated from Savannah State
College in 1963 and then Howard University Law School. A mes-
merizing orator, Hill in 1968 became the first black elected to the
Georgia House of Representatives from Savannah since Reconstruction. In 1971 Hill established a law firm in Savannah and quickly began representing the city's black residents. Working for Hill in the
summer of 1973, Thomas represented black families whose coastal
estates were under threat of commercial and residential development.
The issue was dear to Thomas's heart because the people he was representing were essentially those he grew up with in Liberty County.
Some were fighting to keep property that had been in their families

for generations. Thomas loved the work. At the end of the summer Hill offered Thomas a job after he graduated. Thomas turned the offer down. Fletcher Farrington, Hill's partner, said the twelve-thousand-dollar salary may not have been enough for Thomas. "To do civil rights work you have to make a lot of financial sacrifices," he said. "He might not have been in a position to do that."

Thomas turned twenty-five over the summer of 1973. He was both a husband and a father. He faced college loans to repay after graduation. He was flat broke, and he needed to make money. Beyond that, he wanted to make money. He was tired of never having enough to buy what he wanted. He wanted a fast car and a nice house. Civil rights law was fine, Thomas thought, but it wasn't going to give him the lifestyle he longed for. The real money, he knew, was in corporate law with the big law firms. Despite his practical work with New Haven Legal Assistance and Bobby Hill in Savannah, he figured he could make a strong case for himself as a corporate lawyer.

Thomas wasn't alone in his determination to strike out for more lucrative job opportunities. By 1974, the year he graduated, there was a marked shift in the mood of the law school. The wind-down of the Vietnam War triggered rapid inflation. The economy was struggling to refocus. President Nixon ordered wage and price freezes to halt business and consumer costs that seemed to be spiraling out of control. In contrast to the idealism of the late 1960s, a "grim professionalism" dominated Yale in the 1970s, Kingman Brewster recalled. "What had been the last spoiled generation of the sixties was replaced by a somewhat economically panicky generation of the seventies," he said.[2]

Thomas's game plan, logical to him, was in fact quite optimistic. Law firms in the early 1970s were beginning to open their doors to black associates, but only barely. So-called progressive firms were hiring black associates only one at a time.

Thomas tasted the sobering realities when the large law firms from New York, Washington, and Chicago came around for interviews in the fall of 1973. Thomas told interviewers that he wanted to practice business law, but they talked only about opportunities in civil rights law. He resented the assumption that he could only practice a certain type of law, and he had already heard from black attorneys who felt they were window dressing at large firms. Years later Thomas told a colleague about a friend at a large firm who was given a prominent glass office close to the reception area, where everyone could see him. Thomas's response, recalled Chris Brewster, an attorney who later worked with Thomas, was "I don't want to be that guy."

Although Washington and New York sounded appealing, Thomas actually had his heart set on Atlanta. Close to Savannah and his grandparents, Atlanta had the additional benefit of being a dynamic, cosmopolitan city. In 1973 it elected its first black mayor, Maynard Jackson. Some of the wealthiest and most successful black business-men in the South lived there. There was a spirit of cooperation among white and black Atlantans that existed in few other places in the South.

Thomas applied to Kilpatrick, Cody, Rogers, McClatchey & Regenstein, an established Atlanta firm with a progressive reputation. Louis Regenstein, Jr., one of the senior partners, had represented Martin Luther King, when the Internal Revenue Service targeted the King family for politically motivated audits. Thomas had interviewed with a lawyer from the firm in New Haven in mid-November and was invited to Atlanta for a follow-up interview with the firm's hiring committee the first week of December.

In Atlanta Thomas interviewed with fifteen of the firm's lawyers over two days. Potentially hiring the firm's first full-time black associate, the white lawyers scrutinized Thomas closely. He told his interviewers exactly what they wanted to hear. "He really wanted to work hard as a corporate lawyer," recalled Joe Beck, who interviewed Thomas. "If it had been said by the average interviewee, I would have thought he was kissing up. But he said it in a way that convinced me he really meant it."

Thomas told Beck to forget about the plaintiff law he had done in Savannah. He would bill a lot of hours. "You've come to the right place," said Beck. "We're the sweatshop of the South."

R. Lawrence Ashe, Jr., the Kilpatrick lawyer who shepherded Thomas through the interview process, told him he'd made a strong impression. Thomas left convinced that an offer would soon be forthcoming.

It wasn't. Nearly three weeks elapsed before the firm contacted Thomas—three times as long as it normally took after a "fly-back" interview. Ashe told Thomas that the firm wanted to hire him but couldn't until it heard back from others who had already been offered jobs. Ashe put the blame on the late-November time slot that Yale had given Kilpatrick for interviews. The firm had already made most of its hiring decisions by the time it got around to Yale, Ashe told Thomas, adding that he hoped they would ultimately be able to hire him. In the meantime, rejection letters from other firms were mounting in a fat stack inside Thomas's New Haven apartment.

Ashe told Thomas only half the story. After Thomas left Atlanta, a disagreement erupted over his hiring. The younger members of the hiring team lobbied vigorously to hire Thomas as the firm's first full-time black associate, Ashe said. But the senior members balked. "They wanted the person to be a lock-cinch for success," he said. Despite a Yale law degree, the older, white partners concluded Thomas wasn't a superstar. Yale's pass-fail grading system was partly to blame. Thomas's transcript gave no indication whether he was at the middle or bottom of his class.

Thomas was livid. How could he not get an offer, graduating as he was about to from the best law school in the country? All the hard work, all the long hours, all the sacrifices had been so he could land a good job. And now he'd struck out. He felt utterly mortified and defeated.

Time was now running out. Thomas needed to line up a job. Once again, help came from an unlikely quarter—a young Republican at-

torney general from Missouri, John Danforth, one of the last interviewers to make the rounds.

The meeting between Thomas and Danforth was more a sales job
than a job interview. Danforth believed that any graduate of Yale Law
School could handle the work he had to offer. His challenge was to
convince Thomas that the job was worth taking. Associate attorney
general jobs in Missouri paid less than eleven thousand dollars a year,
so Danforth tried to sell Thomas on the experience. He promised him
he could practice whatever type of law he liked. Danforth said he
would put Thomas in the courtroom to litigate. Thomas told Danforth he wasn't interested in civil rights cases. Fine, said Danforth. "I
will offer you more experience and less pay than anyone else who is
interviewing you," Danforth told him. Intrigued, Thomas agreed to
visit Danforth in Jefferson City, Missouri, where the attorney general's offices were located.

In Jefferson City, Thomas saw the attorney general's offices, a grim
warren of drab spaces located in the basement of the supreme court
building. Jeff City itself was a cow town, a good three hours' drive
from nowhere. All in all, the environment looked about as desolate
and unwelcoming as Thomas could imagine. It was the last place on
earth that he wanted to start his law career. Thomas liked the people
he met, however, and other attorneys in the office talked up the qual-
ity of the experience they were getting. "There's plenty of room at
the top," Danforth told Thomas. "Yeah, right," Thomas muttered
under his breath. Easy for you to say—you're rich, he thought to
himself.[3]

In his meetings with the hiring committee, Thomas repeated that
he did not want responsibilities that might be considered typical for
minorities, such as representing the Missouri Human Rights Commission. He wanted to be treated no differently than if he was white,
he said. Naïvely, he also asked for a heavy caseload. Alexander
Netchvolodoff, Danforth's chief of staff, started laughing. He pulled
out a computerized docket of nearly five thousand pending cases.

"We've got a little overflow here," he joked. "What percentage of this load would you like to take on?"

Having convinced Thomas to visit Jefferson City, Danforth pressed him for a quick decision. Faced with a choice between uncertainty and a sure thing, Thomas took the sure thing. The salary—more than one thousand dollars less than the twelve thousand dollars he turned down from Bobby Hill in Savannah—was perhaps most difficult to swallow. Thomas would be miles and miles from Georgia, *and* he would be making less money. It was humiliating.

In late January or early February of his final semester at Yale, Kilpatrick ultimately came through with a job offer, an associate position that would have paid Thomas fifteen thousand dollars a year. But it was too late. Thomas had given Danforth his word. There was no going back.

Thomas graduated in May 1974. He had worked so hard for this moment. And for what? An eleven-thousand-dollar-a-year job in Jefferson City, Missouri. His classmates were buzzing about some of the lucrative and interesting jobs that awaited them. In his dejection, Thomas thought he saw smirks.

Once again, Anderson wasn't in the audience to see his grandson graduate, refusing to make the trip north. Perhaps Thomas's only solace on that occasion, the day he received his Juris Doctor of Law, was that he had finished without giving in to the temptation to quit. On the whole, however, he had never felt so frustrated in his life.

Back in Atlanta, Kilpatrick lawyers filed Thomas's job application alongside those of others who hadn't made the cut that year. Buried in the file was a short note from the pen of Phillip S. Heiner, one of those who interviewed Thomas. "A real overachiever," wrote Heiner. "Predict he will do well wherever he decides to go."

CHAPTER 8

"NOT BAD FOR A STARVIN' MAN"

Shortly after graduation, Thomas left for Saint Louis to study for the Missouri bar exam. Danforth arranged for him to stay with a local legend, Margaret Bush Wilson, who headed the Saint Louis NAACP chapter and had been deeply involved in the Missouri civil rights movement. Wilson recalled the freshly minted law student as eager and engaging, all southern manners and respect. Yet Thomas's enthusiasm masked lingering bitterness over his rejections by Kilpatrick and other silk-stocking law firms. The rejections badly damaged his self-esteem and contradicted everything he had been taught about the rewards of hard work and discipline. Thomas had believed fiercely in the power of hard work to get what he wanted. He had played by the rules only to see that the rules didn't matter.

In his dejection Thomas also focused some of his anger on affirmative action. Rightly or wrongly, he believed white interviewers assumed he could only have been admitted to Yale and earned his degree because of special breaks. His hard work and effort was tainted, he thought, by the unspoken presumption that a black student could not succeed at a place like Yale without a helping hand. Thomas felt stigmatized, and he blamed affirmative action.

"Clarence and I talked many times about the damage that affirmative action was doing to our reputations," recalled DeShay, who kept in touch with Thomas after he left Yale. "The things that he and I worked so hard for . . . had been trashed, because of the perception in the larger society—and even among too many black folk—that whatever we have, we have because somebody gave us some break."

Other black men and women of Thomas's generation experienced the same queasiness but came to different conclusions. Just a few years after Thomas left Yale Law School, Stephen L. Carter was insulted when apologetic admissions staff at Harvard called to say there had been an "error" in rejecting him from the law school. The school assumed Carter was white when it rejected him. Now, learning that he was black, they laid out the welcome mat. "The insult I felt came from the pain of being reminded so forcefully that I was good enough for a top law school only because I happened to be black," Carter wrote in *The Wall Street Journal* years later. Like Thomas, Carter felt stigmatized by affirmative action. But unlike Thomas, Carter faced the stigma, shrugged his shoulders, and said, "So what?"[1]

For Thomas, however, the stigma cut too close to home. Proving he was the equal of white students had been his motivation to work hard in school, especially at those moments when he felt like giving up. Now, because of policies that he'd had no hand in creating, he was being tagged as some kind of special-needs case. And the hurt fueled his resolve that he would not allow his black skin to define who he was or dictate his career path. Thomas was at the beginning of a lifelong battle to define his own identity in a society ruled by racial assumptions and stereotypes. It was a noble goal, and like most noble goals, easier said than achieved. Indeed, Thomas was unaware that his detour to Missouri had already been dictated by his race. Danforth wanted good lawyers in his office. But he specifically wanted a black lawyer. In Thomas Danforth saw an opportunity to right the wrongs of discrimination, a civic obligation shaped by his own experiences with segregation.

Born in 1936 to Donald and Dorothy Danforth, John Danforth was
the third generation of a Missouri legend. William H. Danforth, his
grandfather, founded the Robinson-Danforth Commission Company
in 1894, a livestock grain-and-feed business. The company survived
tornadoes and near-financial ruin before emerging as Ralston Purina
in the twentieth century. Danforth's father was chiefly responsible for
building the company into a multibillion-dollar business. But it was
Danforth's grandfather who planted many of the moral, religious, and
philanthropic seeds that ultimately shaped Danforth's life.

Ralston Purina created a wealthy lifestyle for young John Danforth. He lived in an exclusive white neighborhood. He attended private Saint Louis schools. The privilege wrapped Danforth in a kind of
destiny. His nickname in school was "the Senator" because it was
widely assumed that he would lead a life of public service.

Every aspect of his life in Saint Louis, however, was segregated. He
did not have black classmates until he reached graduate school. His
most vivid memory of black people in Saint Louis was from attending amateur boxing matches. He can still see himself as a young boy
of perhaps nine or ten, sitting next to his father at a match between a
black and a white boxer. He can hear the crowd mutter in unison,
"Come on, whitey!" Louder still, "Come on, whitey!"

After prep school, Danforth attended Princeton, where he majored
in theology. He temporarily gave up his political ambitions and enrolled in divinity school at Yale. But after two years, he switched to law
school, graduating from Yale with degrees in both law and divinity.

At Yale Danforth spent many weekends visiting his aunt in Scars-
dale, New York. She was passionate about civil rights for black Americans, and she spent many hours talking to her nephew about the ills
of segregation and racial hatred. Following in his family's footsteps,
Danforth became a member of the Danforth Foundation, founded by
his grandfather, and took up the foundation's cause of funding historically black colleges. In the 1960s he was chosen to become a
trustee at Morehouse, serving alongside Martin Luther King, Jr., and

his father. Danforth's association with Morehouse also introduced him to the university's president, Dr. Benjamin Mays. A brilliant man of commanding dignity, Mays was one of the South's leading voices in the nascent civil rights movement. Through Mays, Danforth said, he began to understand how black Americans had suffered from discrimination and a denial of their equal rights as American citizens.

Danforth left Yale in 1963 to join a law firm in New York City. Although he had abandoned the path to the priesthood, an Episcopalian bishop in Saint Louis approached him with an unusual proposition: He wanted to make Danforth a citizen priest. Danforth became a deacon, ministering to the sick at a Manhattan church. When he returned to Saint Louis three years later, he was ordained as an Episcopalian priest, a distinction that later earned Danforth the nickname "Saint Jack" in the U.S. Senate.

In 1968 Danforth ran for Missouri attorney general against a well-entrenched Democratic incumbent. Missouri voters had not elected a Republican to any statewide office in nearly three decades, and Danforth was written off.

Danforth campaigned hard against the Missouri Democratic machine, telling voters he was a fresh, ethical force to end the political spoils system that dominated state government. He also capitalized heavily on the Danforth name and family fortune. He was two months past his thirty-second birthday when he triumphed on election night.

In his new job, Danforth scoured the state and elite law schools around the country for the best lawyers he could find. Some he lured away from big law firms paying much more, captivating them with his determination to clean up government. Danforth did not ask about the politics of his recruits, believing that partisan hiring would breed the spoils system he was trying to root out. Ambitious Republicans, however, were eager to work for Danforth, and the attorney general's office became a springboard to higher office for several of them. Former Missouri governors Christopher "Kit" Bond and John Ashcroft both worked for Danforth before their respective elections in 1972 and 1984. (Ashcroft also succeeded Danforth as attorney general in 1976.) The political hires, however, were the exception in Danforth's office.

Minority recruitment brought Danforth to Yale in the fall of 1973, when he first met Clarence Thomas. "I was looking for a black lawyer," said Danforth. "Politically, I thought that it was important. I didn't want, when asked the question, Well, how many African Americans are working in your office? I didn't want to say none."

After sitting for the Missouri bar exam in Saint Louis, Thomas collected Kathy and young Jamal and brought them to their new home on the banks of the Missouri River. Set atop several low-lying hills at a sharp bend in the river, Jeff City supported a single industry, the Missouri state government. Downtown streets were named after former presidents. Townspeople, many of German ancestry, addressed one another as Mr. and Mrs. The city was also home to Lincoln University, founded by black veterans of the Civil War. The college supported a small community of black faculty, staff, and students. Clarence and Kathy rented a house on the east side of town, near the college.

With the exception of Thomas, Danforth's staff of about thirty lawyers were all white men. Accustomed to being the only black guy, Thomas fell in easily with his new colleagues. The dynamics of the office helped. Most of his new coworkers were in their mid-twenties to mid-thirties, and there was little hierarchy. Danforth's crusading mentality gave everyone a sense of mission. They viewed themselves not as government functionaries but as elite professionals carrying out divine providence. The macho attitude translated into flippant give-and-take. "No one in the office suffered fools," said Harvey Tetlebaum, one of Danforth's first hires. "If you took yourself seriously, you were ridden out of the office."

Thomas relished the office's competitive, irreverent edge, and he made friends quickly. "He wanted to be treated like any other lawyer," said Tettlebaum. "He didn't want any special treatment."

Thomas befriended everyone in the building, including the white janitor who mopped the floors. "He was always the person who knew people and who spoke to them," Danforth remembered.

After work Thomas frequently hit the local bars with his colleagues, unwinding after the ten- to twelve-hour workdays that were routine. A Friday night in early September 1974, just after Thomas arrived, remains a footnote in Missouri legal lore.

Thomas and a half dozen of his office mates took supreme court clerk Tom Simon out for a few rounds at a local bar after work. Almost immediately, the young assistant attorney generals started pestering Simon to show them their bar-exam results, which he was supposed to post the next day. Simon demurred. More rounds of beer followed. Soon it was after midnight. The young lawyers pointed out that it was now officially Saturday: Simon could retrieve the results without breaking the Saturday embargo. Less firm in his resolve by this time, Simon relented, and the group weaved its way back to the court clerk's office. Unable to find the bar-exam results, the well-oiled lawyers began rifling through every filing cabinet in his office. At last someone found the likely cabinet, but it was locked. A search for the key was fruitless. Someone, however, found a screwdriver.

"Next thing I know," recalled Simon, "someone says, 'Hey, it's open!' "

To his relief, Thomas learned that he passed the bar exam. He was admitted to the Missouri bar on September 13, 1974, taking his oath before members of the Missouri Supreme Court. Missouri would remain the only state where Thomas was licensed to practice law.

Following office practice, Thomas was first assigned to the criminal-appeals division, which briefed all the petitions to the appeals and supreme courts. True to Danforth's word, Thomas was arguing appeals before both courts within weeks of being admitted to the bar.

The caseload was staggering. Thomas developed a routine of getting to his desk at four-fifteen in the morning before brewing a pot of coffee and sitting down to work. Danforth periodically popped his head inside Thomas's tiny office later in the day to ask how he was doing. "Not bad, for a starvin' man," was Thomas's usual reply, a dig at Danforth's low wages.

After seven months Thomas moved from criminal appeals to the revenue-and-taxation division. He was interested in the work but also saw it as the most promising route to a lucrative job in private practice. He viewed his stint in Missouri as only a way station toward his ultimate goal of making real money. The hunger for a fat salary expressed itself in Thomas's office decorations. He hung a photograph of a Rolls-Royce above his desk. He told his colleagues what he had said to his father-in-law: One day he'd be able to afford a car like that.

While many lawyers in the office recoiled over arguing and briefing revenue cases, most of which involved some of driest and most arcane aspects of the law, Thomas reveled in their intricate details. Most involved questions of sales tax and revenue collection. What animated Thomas most in those cases, colleagues recall, was a zealous regard for fairness. In one case Thomas went after a class of trucks that were dodging sales taxes because of their vehicle classification. In another he defended the state's right to collect sales tax from bowling-alley operations, which had escaped paying tax in an administrative loophole. In a case that attracted some local publicity, Thomas defended the state's plan to phase out low-numbered license plates, which Missouri governors had handed out as political favors. Even Danforth worried that Thomas might ruffle too many feathers. Plate holders regarded them as family heirlooms and badges of prestige. But Danforth left the decision to Thomas, who won the case on appeal.

Thomas loved the independence. "Nobody ever supervised me," he recalled. "I was my own man." After years of living day in, day out under crushing racial stereotypes, Thomas experienced a lightness and freedom that he had not known. "It wasn't black this and black that,"

Thomas said. "I didn't have to play any roles. Our time wasn't concerned with the overarching racial issues. You got to be you."

One of Thomas's closest friends was a man who, on paper, could not have been more different from him. Eight years older than Thomas, Richard Wieler was born and raised on a farm in eastern Nebraska. At the age of fifteen he came down with polio. A late diagnosis nearly cost him his life. He spent six weeks in an iron lung and a year on assisted-breathing devices. When he went home, he was completely paralyzed from the neck down. Wieler's sister wasn't so lucky, dying of polio just days before her twelfth birthday. For a long time, Wieler's parents were convinced their only son would meet the same fate; indeed, doctors had counseled them to read him poetry to ease his final days.

Rehabilitation at Warm Springs, Georgia, where President Franklin Roosevelt underwent polio treatment, helped Wieler regain extremely limited use of one hand, and he was fitted with a manual wheelchair. In 1959 an uncle helped convince administrators at the University of Missouri that Wieler could handle college. The school was not configured for wheelchairs, and Wieler was constantly having to arrange to have himself carried up and down stairs to get to class. In Warm Springs Wieler had been taught how to type using a long stick that he held in his mouth, and he typed his term papers one key at a time. He graduated from the University of Missouri Law School with honors.

At first Thomas felt extremely uncomfortable around Wieler. He'd never known someone with such a severe disability. He himself was so strong and athletic; Wieler, by contrast, seemed weak and helpless. Yet an odd friendship developed between them. Just as Thomas battled racial assumptions, Wieler was constantly battling assumptions about his disability. Like Thomas, Wieler wanted respect. He wanted to be regarded for his legal abilities, not his handicap. Thomas came to admire Wieler's fierce independence and how he never complained about his disabilities. Wieler's struggles put Thomas's in perspective. How could he complain about how he'd been treated when Wieler's struggles began every morning when he got out of bed?

In 1975 a colleague gave Thomas a *Wall Street Journal* article about economist Thomas Sowell, whose book Thomas had dumped in the trash three years earlier. Now, far removed from Yale's liberal political environment, Thomas was suddenly curious about Sowell. After reading the review, Thomas ran out and picked up his latest book, *Race and Economics.*

Sowell was an intriguing personality. Raised in North Carolina, he moved to Harlem as a young boy. He dropped out of tenth grade to work before being drafted into the Army in 1951. After discharge, he attended night classes at Howard University on the GI bill while working as a stock clerk in a photo store. He transferred to Harvard and graduated magna cum laude with a degree in economics. He earned his economics doctorate at the University of Chicago, where he studied under the noted conservative economist Milton Friedman. Sowell's academic brilliance was dazzling, yet he was an iconoclast, and his quirky temperament left colleagues either endeared or enraged.

In *Race and Economics* Sowell attempted to distinguish patterns of economic achievements and setbacks among a wide range of ethnic minorities, including black Americans. He examined the progress of Jewish and Irish immigrants to the United States, early Japanese settlers on the West Coast, and Puerto Rican immigrants to New York City. He attempted to identify the characteristics that had allowed each group to climb out of poverty and achieve wealth in America. He found success where ethnic cultures had emphasized self-reliance, education, and hard work. These groups, Sowell noted, often succeed despite widespread racial prejudice against them.

Sowell devoted particular attention to the economic history of black Americans, beginning with the progress of freed slaves prior to the Civil War. He found steady economic advancement among them—even in the face of harsh legal restrictions on free men. Sowell concluded that discrimination, prejudice, and hostile laws were often overridden by determination, skills, and resourcefulness, along with the prevailing headwinds of the economy. He also concluded that ethnic groups owed little of their economic success to political enfranchisement. The thrust of Sowell's research concerning black Americans, however, was that their strength lay in their resilience and self-reliance in the face of hostile laws and white bigotry. "If the history of American ethnic groups shows anything," Sowell wrote, "it is how large a role has been played by attitudes—and particularly attitudes of self-reliance."[2]

No other writer was to have as much influence on Thomas's thinking about racial policies in America. "Finally, here was a black guy applying numbers and facts and figures to people's assertions," said Thomas. "What I wanted was to apply systematic thinking to problems that were important to me, and here was a guy doing it, and who wasn't caught up in this almost religious fervor about it." Later, when Thomas learned that the author was speaking in Saint Louis, he showed up with one of Sowell's books for him to autograph. Thomas's enthusiasm penetrated Sowell's normally prickly personality, and the two developed a friendship that extended for three decades.

Sowell directed Thomas into a fresh round of soul searching. Thomas began thinking not about the forces that kept blacks back— the focus of his energy at Holy Cross—but about what had made them successful.

He began by examining his own family, particularly his grandparents. Anderson succeeded because of his thrift, industry, and hard work, Thomas concluded. Thomas had succeeded because Anderson had saved and sent him to private school, giving him an opportunity that Anderson himself had been denied. Like Sowell, Thomas saw that Anderson's self-reliance and dedication to the betterment of his family were what had helped the family succeed.

Thomas's burgeoning belief in economic self-reliance conflicted with social policies being advocated by civil rights leaders. To improve education for black Americans, civil rights leaders were aggressively promoting busing. To solve the problems of poverty, they pushed for increased federal spending for social-welfare programs. Thomas found himself increasingly at odds with both approaches.

As the first black to attend an all-white school in Savannah, Thomas had firsthand experience with integration. The encounter nearly shattered his self-confidence as a teenager. How would even younger blacks fare when put with whites who considered them inferior? Given the racial animosity still seething in America, Thomas became convinced that busing black children would be disastrous for their education—not because they were inferior but because they would be made to feel that way. He also bristled at the implication that anything "white" had to be better. The segregated Catholic schools of Savannah may not have had every resource available, but there was no telling Thomas that they were inferior simply because they were all black.

Welfare policies insulted Thomas's concept of self-reliance. Anderson became livid when Thomas's sister, Emma Mae, went on public assistance in Savannah after the birth of her four children. That Emma Mae was a single parent seemed not to sway Anderson; Thomas also took an excessively hostile view of his sister's reliance on welfare. At one point the two were not on speaking terms because of it, according to Holy Cross classmate Clifford Hardwick.

Like Anderson, Thomas saw welfare as a crutch that was slowly breaking the self-reliant ethos of the black community. Worse still, he believed the welfare laws, as they were first constructed in the 1960s and 1970s, discouraged fathers from living at home, driving families apart and pushing single parents into poverty.

Thomas found validation for his beliefs on sporadic trips home to Savannah, where urban renewal projects had cleared out run-down tenements and replaced them with even grimmer housing projects. Drug dealers were creeping into the neighborhood. Anderson and the older neighbors complained that they no longer felt safe.

More and more, Thomas came to see the Great Society programs of the 1960s and 1970s as well-meaning but misguided experiments on black Americans. Underlying the concern was a deep-seated belief that government was not and could never be a constructive force in their lives. The government had legalized segregation and discrimination, enforcing the laws and rules that kept black people down. Everyone in the black community of Thomas's youth, especially Anderson, viewed government as the enemy. Thomas never rid himself of the same suspicions.

Thomas's emerging conservatism foreshadowed later clashes with civil rights leaders. In private, Thomas already relished taking shots at established civil rights icons such as Jesse Jackson and other "so-called black leaders," recalled Larry Thompson, a fellow black conservative who befriended Thomas in Saint Louis and later became deputy attorney general under President George W. Bush.

Thomas became "apoplectic" over the idea that all black people should think alike, recalled Thompson, who also grew up in a blue-collar home. Both men were equally frustrated by how civil rights leaders attempted to marginalize people who didn't think as they did. Class distinctions animated their thinking.

"We both had a sense that some black people, who are our contemporaries, who were the sons and daughters of middle-class black people, were sort of playing the system," said Thompson. "Things like affirmative action really didn't apply to people like Clarence and I as much as it would to the son or daughter of a doctor or a lawyer. They would be the beneficiary of something like that, not us." Some of this thinking was more perception than reality. Thomas, of course, had benefited from affirmative action.

Thomas, Thompson recalled, was constantly arguing against government programs that sapped black Americans of their independence. "We needed to form strong work habits, strong institutions, and not be seduced by the largesse of having the government help us do everything," said Thompson, summarizing Thomas's beliefs at the time.

In Danforth's office, Thomas became more outspoken about his conservative beliefs, surprising Danforth with his zeal. "We were stunned, absolutely stunned, when we discovered that we had hired someone who had a conservative political philosophy," recalled Alexander Netchvolodoff, the former chief of staff. To this day Danforth recalls Thomas as the most conservative member of his staff, including, at the time, John Ashcroft.

Ironically, in Danforth's office Ashcroft and Thomas were not particularly close. The two men clashed repeatedly both on and off the job. As a fallen-away Catholic, Thomas delighted in goading Ashcroft over his religious beliefs. Deeply devout, Ashcroft was constantly quoting scripture to Thomas, whose own thorough knowledge of the Bible was shaped by years of training for the priesthood. "Whatever John quoted to Clarence, Clarence could come up with the Bible verse that said exactly the opposite," recalled Russ Still.

Thomas and Ashcroft also butted egos on the basketball court, where Ashcroft's physical play was notorious.

Years later Thomas said he apologized to Ashcroft for mocking his religious beliefs. In 2001 Ashcroft chose his former Missouri officemate to swear him in as United States attorney general.

Thomas also enjoyed mixing it up with black students at Lincoln, where Kathy was enrolled in school. He chided students for filling up their curriculums with African-American studies instead of more fundamental courses like English, science, and math. He told them that he had been an English major because English was a skill that would help him get ahead.

In 1976 Danforth ran for United States Senate and won, leaving the attorney general's office in early 1977 to move to Washington.

Danforth's departure was a good opportunity for Thomas to move on, and Thomas considered a number of different options. Danforth invited him to join his staff in Washington, but Thomas declined. Around the same time, Thomas also had a heart-to-heart talk with John Bardgett, Sr., then a Missouri Supreme Court justice. Bardgett, a

Democrat, liked Thomas and was impressed by his legal skills before the court. Had Thomas ever considered becoming a judge? he asked. Bardgett was politically well connected and told Thomas that he probably could help him get an appointment to a Missouri state court in Saint Louis. Like Danforth, Bardgett believed the Missouri bench, overwhelmingly white, needed greater diversity. Thomas thanked him but said no.

Now twenty-eight, Thomas was ready to make some money. Danforth had connections at Monsanto, an agricultural conglomerate based in Saint Louis, and Thomas was offered a job in the company's legal department. It was time, Thomas told Harvey Tettlebaum, to pay off some student loans. He said his good-byes and moved to Saint Louis that year.

In the end the decision to work for Danforth, made in bitterness and desperation two years earlier, had had a positive result. Thomas had proven his abilities as a lawyer and made good friends. More significantly, he'd been free to be himself, to live life without feeling as if people were always judging him because of his skin color. "It was the best job I've ever had," said Thomas, sitting in his Supreme Court chamber more than twenty-five years later. "Still, to this day, it's the best job I've ever had."

CHAPTER 9

BLACK CONSERVATIVE

Clarence Thomas stood bolt upright in the small conference room, smiling nervously and surveying the invited guests through a pair of oversize, horn-rimmed spectacles. The Afro from his Holy Cross years was gone, replaced by a short crop of black hair. Only a neatly trimmed goatee hinted at the heady radicalism of his twenties. A month short of his thirty-fourth birthday, Thomas was about to be sworn in as chairman of the United States Equal Employment Opportunity Commission. Beads of sweat trickled from his forehead and temples, ringing his starched white shirt with perspiration. Connecticut Supreme Court justice Angelo G. Santaniello, who'd helped Thomas find a job in New Haven eleven years earlier, stood opposite him to administer the oath of office. Thomas's mother, Leola, stood between the two men, holding a family Bible. In his awkwardness, Thomas put his right instead of his left hand on the Bible, only to be gently corrected by Santaniello.

After swearing allegiance to the Constitution, Thomas turned toward the audience to make his remarks. He chose a Robert Frost poem to capture the moment, explaining that he had often read the poem in high school to cope with the difficulty of being the only black student in an all-white school. " 'Two roads diverged in the woods,' " Thomas said. " 'And I, I took the one less traveled by.' "

Not only less traveled, Thomas's road had also been surprisingly short. Just two and a half years after arriving in Washington, he was already on his second appointment in the administration of President Ronald Reagan. The success, however, came at a price Thomas once vowed never to pay: abandoning his determination not to take jobs tied to his race.

The erosion of principle began with a fateful decision to leave Missouri and join Danforth in Washington after two and a half years at Monsanto. Thomas liked corporate legal work well enough but had become increasingly frustrated by his lack of advancement at Monsanto, according to his father-in-law. He wanted a clear indication that the company planned to move him up the line, but he wasn't getting one. Danforth's invitation caught him at just the right moment. For the fourth time in the eight years since he began law school, Thomas packed his family and moved, leaving Saint Louis and the only private sector job he would ever hold.

Arriving in Washington in August 1979, Thomas was convinced that his new home would be yet another way station toward his goal of practicing law in Georgia. He left books packed in boxes in anticipation of another move. Thomas and Kathy rented a house outside the Beltway in Bethesda, Maryland, just north of the Bethesda Naval Hospital. The house was an expensive wreck, and Thomas spent weekends pulling up rotting carpet and repainting the walls.

Once again, Thomas told Danforth that he didn't want assignments coded to his race. On Capitol Hill Danforth assigned him to handle energy issues on the Commerce Committee. The Middle Eastern oil embargo made America's dependence on foreign oil a hot issue.

As both black and now identifiably Republican, Thomas quickly learned that he was an oddity in Washington. Only a handful of black Republicans worked on Capitol Hill. "We were basically treated like we were from Mars," recalled Phyllis Berry Myers, who worked at the Republican National Committee for then-chairman William Brock.

Even before Reagan's election in 1980, the Republican Party

found it extremely difficult to attract black voters to the party of Lincoln. Barry Goldwater, the party's 1964 presidential candidate, alienated black voters by appealing to the racist passions of southern whites, many of whom were bitter over integration. By the late 1970s, most Republicans took for granted that black Americans voted Democratic. "The damage was very deep," said Brock. "It took some very, very courageous people to say, Yes, I will identify with the Republican Party."

In December 1979 someone dropped a copy of the *Lincoln Review* on Thomas's desk. The newsletter featured conservative opinions by black Americans. Thomas devoured the issue and called the newsletter's editor, J. A. "Jay" Parker, who invited Thomas to lunch the following week.

The lunch launched an important friendship for Thomas, both personally and professionally. Parker was older than Thomas, old enough to be his father, and perhaps Washington's best-known black conservative. His mother had been a maid in his native Philadelphia and his father a short-order cook. Parker began working when he was nine, scrubbing stoops and delivering newspapers. He joined the Republican Party in 1955 and worked on Goldwater's presidential campaign. Later he served on the board of Young Americans for Freedom, whose college chapters supported the Vietnam War and were virulently anticommunist.

Parker was a libertarian and arguably more conservative than Thomas himself. He opposed virtually all government regulation and was deeply hostile toward the Great Society programs of the Johnson administration. He was the first black American Thomas met who not only wasn't afraid to proclaim his conservative beliefs but delighted in broadcasting them. Thomas admired Parker's tenacity and the fact that he didn't particularly care what people said about him.

The two men also discovered a remarkable coincidence. Parker had briefly attended Haven Homes Elementary School in Savannah a decade before Thomas.

Parker and his wife took Thomas under their wing, inviting him to their Virginia home for dinners and weekend afternoons, and treating him like family. Thomas looked up to Parker almost like a father, the one role model that had been completely absent from his own life.

The year 1980 triggered an earthquake in the nation's political landscape. Ronald Reagan won the presidency in a crushing landslide over President Jimmy Carter. Six years earlier the GOP had been badly tainted by the Watergate scandals and the near-impeachment of President Nixon. Reagan rebuilt the GOP coalition with a strong appeal to white, conservative Democrats and working-class families. He campaigned hardly at all for black votes.

But for Clarence Thomas and a tiny band of black conservatives, Reagan's victory signaled new opportunity. For Thomas, Reagan's victory also marked the point when his professional and personal lives began to move in opposite directions.

A month after Reagan's election, a small group of black conservatives led by Thomas Sowell, then a scholar at Stanford University's Hoover Institute, organized a two-day conference in San Francisco. Calling it the Black Alternatives Conference, organizers hoped to map out a new political agenda for black Americans. Sowell invited Edwin Meese III, one of Reagan's top campaign advisors, who was certain to play a prominent role in the new administration. Because of Sowell's work in academia, he had contacts all around the country, and not just among conservatives. Those who responded to invitations included Randolph W. Bromery, the first black chancellor of the University of Massachusetts; Charles Hamilton, the Columbia University political scientist who coauthored books on the black power movement with Stokely Carmichael; and Percy Sutton, the former borough president of Manhattan and legal advisor to Malcolm X. Conservatives at the conference included Clarence Pendleton, a friend of Meese's from San Diego who would later be named chairman of the U.S. Commission on Civil Rights.

Walter Williams, an economist like Sowell, was also at the conference. At the time Williams was working on his book *The State Against Blacks*, which examined how government regulations fell particularly hard on black Americans. Williams's research dovetailed closely with Sowell's work on ethnic minorities in America. Sowell and Williams were attempting to use economic analysis to show that discrimination alone did not explain the enormous gaps between white and black America. They argued, for example, that minimum wage laws discouraged employers from hiring inexperienced and unskilled workers, resulting in fewer opportunities for black teenagers, and that occupational licensing laws often kept black tradesmen out of the economy because they favored white workers who'd had the benefit of more education and training. "Nobody was denying the existence of racial discrimination, but racial discrimination doesn't explain all that it purports to explain," said Williams.

Thomas looked forward to the conference with the eagerness of a schoolboy. All his conservative heroes were planning to be there, and he believed, naïvely it would turn out, that the potential of the conference was boundless. For five years Thomas had been chafing under the shadows of liberal black leaders who commanded so much of the nation's attention and who were so closely allied with the Carter administration. Conservatives were finally getting their turn, Thomas thought, and he couldn't wait to sink his teeth into substantive debates on education and economic policy.

In the offices of *The Washington Post*, meanwhile, editorial-page editor Meg Greenfield was trying to convince Juan Williams, one of her young black editorial writers, to jump on a plane and attend the conference. Greenfield knew Sowell and thought he was the most provocative thinker among black conservatives. She was also certain that he would play a prominent role in the new Reagan administration. The conference could be a good, informal way to get to know Sowell, she told Williams. Why not go to San Francisco and write a column about him? Williams agreed.

Out in San Francisco, however, the twenty-six-year-old writer found Sowell aloof and prickly. Just trying to get him to answer a question was an act of futility, Williams recalled. Over lunch Williams struck up a conversation with a young black man sitting near Sowell. He was warm and engaging, interrupting himself with frequent guffaws. The contrast to Sowell was as stark as it was refreshing. Forget Sowell, Williams thought to himself; here was the story. He introduced himself and shook hands with Clarence Thomas.

"Thomas stood out from this group because he was genuinely excited about ideas and he wasn't afraid to speak his mind," recalled Williams. "He didn't really care about the power, he cared about the ideas. He was just an engaging person. Here I am struggling as a reporter to penetrate Thomas Sowell, and finding little if any water in that mine; then, holy smokes, here's this guy who's just gushing with energy and ideas."

After lunch the two young men took a long walk and continued to talk. Thomas was brutally candid about himself and his politics. Williams had his story. Several days later the *Post* published the piece under the headline BLACK CONSERVATIVES, CENTER STAGE. The story began: "You've heard about Clarence Thomas, but not by name. He is one of the black people now on center stage in American politics: he is a Republican, a long-time supporter of Ronald Reagan, opposed to the minimum wage law, rent control, busing and affirmative action."

In the span of a few sentences Thomas went from unknown black conservative to vanguard of a new political movement in Washington. The article put him on the map and brought him one step closer to his first appointment in the new Reagan administration.

Two sections of Williams's article would dog Thomas for years to come. Williams related Thomas's adamant resolve not to take jobs in any way relating to black issues, and quoted his explanation: "If I ever went to work for the EEOC or did anything directly connected with blacks, my career would be irreparably ruined," Thomas said. "The monkey would be on my back again to prove that I didn't have the job because I'm black. People meeting me for the first time would automatically dismiss my thinking as second-rate."

Farther down in the story, Williams quoted Thomas on his opposition to welfare and its impact on his own sister, Emma Mae. "She gets mad when the mailman is late with her welfare check," Thomas said. "That is how dependent she is. What's worse is that now her kids feel entitled to the check too. They have no motivation for doing better or getting out of that situation."[1]

Thomas was mortified when he saw the remarks about his sister in print. He believed that he had just been talking to Williams and didn't fully comprehend that anything he said might be quoted. In print, the comments looked as bad to Thomas as they did to everyone else. "I was horrified," he said. "I went home that Christmas. I went home and talked to her, and she said, It's true. It does have these negative effects. She agreed with me. I still didn't feel better. I was mollified but didn't feel that much better. . . . Seeing it in print was horrifying because I didn't mean to do her any harm. So I had to live with that."

Thomas would also soon eat his words about working on black issues. To Williams they seemed less an indictment than a courageous declaration of independence. Thomas struck Williams as an almost heroic figure, unafraid to stand up to stereotypes that weighed down on black Americans, including himself.

"I immediately was gripped by the power and honesty of what he was saying," said Williams. "That if he was dealing in civil rights, that if he couldn't have done taxes for Danforth, if he wasn't talking about issues beyond civil rights, then everyone says, 'Well, either he's here because he's black or that's all he can do.' The typical putdown that white people put on black people.

"He was sort of a fresh face, an honest and a young man who sees his world as his oyster at that moment and feels free to say indeed who he wants to be. It's a rare moment for a black person in America. And he felt free. He was, through my reporter's eyes, the energizing figure at that conference."

Reagan was sworn in as the nation's fortieth president in January 1981. At the White House Pendleton James, his personnel director, faced the daunting task of filling thousands of appointments to government agencies. Among the most difficult jobs to fill were at the Department of Education. Reagan believed education was the province of local government and publicly questioned the purpose of a federal bureaucracy devoted to it.. "Nobody wanted to work for Ronald Reagan at the Department of Education," said James. "He wanted to close the place down."

The administration also faced a crisis over the scarcity of minorities who were willing to work for it. At a certain moment, James recalled, he issued a directive to his staff: "Quit bringing in all these white guys." The White House sent out urgent pleas on Capitol Hill and to contacts around the country for the names of blacks who might be willing to serve in the administration. Many potential candidates simply said no, thank you, James said. "We got turned down quite a bit in the African-American community. . . . They said, 'No, no, I can't leave my family or I can't afford it.' But you knew they just didn't want to. The African-American community hated Ronald Reagan. They didn't dislike him. They hated him."

James doesn't know who recommended Clarence Thomas to him that spring. Netchvolodoff, Danforth's then–chief of staff, believes it was someone on Senator Jesse Helms's staff but can't be sure. James didn't care. "Let me see him," he told his staff.

James summoned Thomas to the White House for an interview. The meeting was cordial and relaxed until James told Thomas that he wanted him to become the new assistant secretary for civil rights at the Department of Education. Thomas's demeanor changed instantly, James recalled. Barely concealing his anger, he looked James in the eye and said, "I am an attorney. I want to work for the Ronald Reagan administration. I am committed to his programs, but I do not want a civil rights job."

James couldn't believe what he was hearing. "I said, 'Look, Clarence, I'll make a deal with you. If you'll take this position, I'll make you a commitment that within six months or eight months, I will come back to you and take you out of that position and put you into, quote, a real job." James thought he had a commitment from Thomas to take the job. Thomas, however, continued to brood over the decision. He'd repeatedly said he wouldn't take a civil rights job. How could he take the job without standing against every principle he believed in? The more he thought about it, the more he concluded that he simply couldn't do it. He told Netchvolodoff that he planned to turn it down.

Stunned, Netchvolodoff tried to talk Thomas out of it. So did other staff members. Finally, Netchvolodoff called in Danforth. "He called Clarence in and said, 'You're going to take this job,'" recalled Netchvolodoff. "I don't care whether you think it's inappropriate and an Uncle Tom assignment, you are going to take it, because this is a stepping-stone that can lead to very great things for you, and you are going to do it."

Thomas turned to one more source of advice: his mentor, Jay Parker. Parker was even more blunt: Put up or shut up.

Thomas took the job, winning quick confirmation before the Senate Labor and Human Resources Committee on June 19, 1981. At the age of thirty-two, he was head of an agency with three hundred people and a multimillion-dollar budget. Larry Thompson, then living in Atlanta, was in disbelief. He wrote Thomas a note, telling his friend he would never doubt his abilities again.

Ironically, the biggest break of Thomas's career also coincided with the biggest personal crisis of his young life. After ten years of marriage, he was separating from Kathy.

"They married very young," said Janet Brown, who worked with Thomas in Danforth's office. "They just grew in different directions."

Thomas moved out of the house and into the apartment of his friend, Gil Hardy, sleeping on the sofa.

In late spring or early summer, sometime before he began his new job, Thomas met Anita Hill. An associate at Wald, Harkrader & Ross, where Hardy worked, Hill also lived in the same Washington apartment building. Like Thomas, she was a graduate of Yale Law School, class of 1980. Like Thomas, she was raised in a hardworking, deeply religious family, growing up on a farm in rural Oklahoma. She attended public schools and, like Thomas, was a bookworm, studying hard to put herself through college and law school. Unlike Thomas, Hill was a registered Democrat who'd voted for Jimmy Carter in 1976 and mistrusted President Reagan in 1980.

Hill said her first meeting with Thomas occurred at a small gathering in Hardy's apartment, a few floors below her own. "He struck me as sincere, if a little brusque and unpolished," Hill later recalled in her memoirs. The two debated some of Reagan's proposed tax cuts and discussed Thomas's forthcoming appointment. Despite the disagreements, they shared a strong bond with Hardy, who had been largely responsible for luring Hill to Wald, Harkrader after law school.[2]

According to Thomas, Hardy told him that Hill now was having problems at Wald, Harkrader and wanted out. Could he help?

Thomas agreed to interview Hill for a job at the Department of Education. In the interview, the twenty-five-year-old lawyer told Thomas that her supervisor at the firm was sexually harassing her, according to Thomas. She'd rebuffed the supervisor's advances but now found that he was giving her poor evaluations and cutting off interesting work assignments, Thomas said. Thomas offered her a job as a special assistant at the Department of Education, reporting to him. In her own account, fifteen years later, Hill does not mention any harassment at the firm. She said her inability to obtain good work assignments stemmed from her failure to carve out a specialty at the firm, and the reluctance of some partners to work with a black, female associate.[3] Hill started work at the Department of Education in August, shortly after Thomas himself began his new job. Although queasy about working in the Reagan administration, Hill said she was excited about the chance to work on civil rights issues.[4]

Thomas stayed less than a year at the Education Department but generated controversy nonetheless over his refusal to enforce changes that he believed would lead to the closure of historically black colleges in southern university systems. Nine years later Thomas's eleven months at Education became controversial for another reason: his alleged treatment of Hill.

Separated from Kathy, Thomas rented a small Washington apartment. Kathy kept custody of Jamal, who came to visit his father on weekends. Outside the office, Thomas plugged in to a network of black lawyers in Washington, some of whom were classmates from Holy Cross and Yale. Some Holy Cross friends were surprised to learn how conservative Thomas had become. Thomas talked politics constantly, said Eddie Jenkins, often arguing that Jenkins and other old friends ought to follow him to the Republican Party. "You'd be a Republican, too, if you had any money," Thomas kidded Jenkins at one Washington get-together. "The only reason there aren't more black Republicans," Thomas said, "is that they ain't got no damn money."

Jenkins said he enjoyed mixing it up with Thomas over politics but increasingly found that Thomas was talking at him, not to him. "When it came to political stuff, it wasn't a discussion," said Jenkins. "It was, 'This is the way it is.' "

Thomas also boasted to Jenkins that GOP power brokers were putting him on the fast track. "I'm their guy," Thomas told Jenkins. "They are moving me."

Thomas's predictions were realized when Pendleton James called him with another administration job fewer than eight months into his stint at Education. But the job, chairman of the Equal Employment Opportunity Commission, wasn't exactly the "real job" that he had promised. Reagan's reputation among black Americans, however, had worsened, and James felt he had few options. The administration had been trying for months to fill the EEOC job and thought it had a candidate in William Bell, who ran a one-man employment agency in Detroit. Bell withdrew his name from consideration, however, when media outlets reported that his firm hadn't placed a single client the previous year. Senate Republicans hadn't the stomach to confirm a nominee with so little experience.

James laid out the situation to Thomas. He begged him to put aside his pride and take the promotion. Thomas balked.

With no other recourse, James told Thomas that he had discussed his appointment with Reagan, and that the president of the United States wanted him to take the job. This time Thomas couldn't say no. But he went to EEOC "kicking and screaming," James recalled. Thomas said he extracted a promise before he took the job: He wanted complete autonomy and no interference from the White House. James agreed but left government service shortly afterward.

Thomas said he was ambivalent about offering Hill a job at EEOC because he desperately needed financial managers, not policy wonks. "She insisted on going to EEOC, and lobbied for it," Thomas said. Thomas said Hardy also lobbied on Hill's behalf. "Finally, I just relinquished, I relented," he said. Thomas offered Hill the option of joining him at EEOC or remaining at Education to work for Harry Singleton, Thomas's replacement and his friend from Yale Law School. Thomas won Senate confirmation to the EEOC on March 31, 1982, and left for the agency in May.

After considering her options, Hill went with him.

CHAPTER 10

THE DUNGEON

wo months after he was sworn in, Thomas trooped down to the EEOC's lobby to deliver his first public remarks to agency staff. Advance reports of the new boss were not terribly flattering. Everyone knew that Thomas was chosen out of desperation when Reagan's first nominee flamed out in the Senate. Thomas had only just turned thirty-four, a kid. He was black and conservative, an unknown species in Washington. And he had already announced plans for an agency reorganization, unnerving employees.

Perhaps sensing the evident skepticism, Thomas opened by declaring that employees might ask some questions afterward. "I emphasize the word *some*," he said.

"I was seventeen years old when EEOC came into being," he began. "And as I matured in intellectual growth and development, I often dreamed of someday assisting in the effort to achieve equality in this society. So my presence here today is a godsend. My reflections upon the treatment I experienced growing up as a child, subjected to the blatant discrimination that existed in so much of the South, along with more subtle discriminatory encounters that I have had to contend with as an adult have led me to draw one simple, fundamental conclusion: This agency is needed. One comprehends quickly that discriminatory practices are not only unfair, unjust, and un-Christian, but also un-American."

Then he stiffened as he launched into a frank characterization of what employees might expect of him. "I assure you, however, no amount of badgering, distortions, or intimidation will change my mind," Thomas said. "I will not go along with popular opinion to get along with people. Popularity does great wonders for the ego but does nothing to define and address seemingly intractable problems we face today."

He glanced around the room. "I accept no one's opinions or views as gospel," he added. "I have no intention of sacrificing my principles to accommodate others, or because it would be more expedient."

He finished and looked up from the podium. No one asked any questions.

Thomas's first speech to employees bluntly displayed the two personalities of the new EEOC chairman: pleading for acceptance but demanding it only on his terms. The stubborn defiance became the hallmark of Thomas's public persona, frustrating friends and enemies alike. It was his suit of armor in a hostile political environment, one that he wore routinely to express unpopular opinions or ward off blows from angry members of Congress. In private, Thomas was a different man, captivating staff with his warmth and infectious sense of humor. The demeanor earned him scores of devoted followers who rallied to defend him when he needed them most.

Thomas served eight years at the EEOC, the longest tenure of any chairman in the agency's history. He fundamentally redefined the EEOC's mission, poured millions into training, and transformed the agency into the respected bureaucracy that it is today. Over his eight-year term Thomas sharpened his views on affirmative action and other race-based approaches to redressing discrimination. He wrapped a constitutional and legal framework around his conservative political philosophy. In his personal life during this period, Thomas divorced, became a single parent, and remarried.

Thomas's personal growth and development, at times bitter and painful, played out against a much larger drama unfolding in Washington and the nation. The EEOC thrust him to the epicenter of a titanic battle over how the nation defined and implemented fundamental notions of equality, opportunity, and nondiscrimination.

Thomas's conservative views won him powerful allies and supporters, who advanced his interests to also advance their own. Those same conservative views, however, flung him into the jaws of equally powerful adversaries, who saw him as a threat to some of the nation's most fundamental civil rights policies. At the EEOC he learned that the third rail of American politics was not Social Security but race. As EEOC chairman, he grabbed the electric current with both hands. The heat burned him badly and stoked opposition to his Supreme Court nomination nine years later.

Thomas was a junior in high school when the EEOC opened for business in 1965. An outgrowth of the historic Civil Rights Act of 1964, the agency was created to enforce the act's provisions barring discrimination based on race, color, religion, or national origin. Reflecting Congress's ambivalence, however, it had no real enforcement powers and was only authorized to encourage nondiscrimination in the work place. A five-member commission appointed by the president oversaw a tiny budget. The agency received no substantive enforcement powers until 1972 when Congress gave it the power to sue employers for discrimination. Even then the law required that the EEOC resort to lawsuits only after mediation failed.

During the 1970s the agency litigated dozens of high-profile class-action lawsuits against major American companies, winning millions in back pay for victims of discrimination. In 1979 Congress gave EEOC responsibility for enforcing age-bias cases and those filed by disabled workers and women charging wage discrimination. By the time Thomas was named chairman, the EEOC employed nearly 3,200 people in fifty offices around the country and had an annual budget of nearly $141 million.

Internally, however, it was an administrative shambles. Unreconciled accounts topped $30 million in 1981. Nearly a third of the outstanding obligations were three years old. Suppliers demanded payment up front because the agency didn't always pay its bills. "Our accounting reports were not worth the paper they were printed on," said Willie King, who worked in the finance department.[1] Auditors discovered $1.2 million in unaccounted-for travel advances, some to employees who had long since left the agency. The depth and scope of the problems surfaced when Cathie Shattuck, Thomas's immediate predecessor, asked the budget officer to show her the agency's balance sheet. He refused. And with good reason. The books had not been closed in over five years, she learned.

Shattuck ultimately found the unreconciled books—literally hundreds of thick volumes—piled inside a large closet in the finance department. Some stacks stretched to the ceiling; others had toppled into heaps. The accounts were not even computerized; finance staff worked with calculators. Photographs of the disarray later became part of a scathing report on the agency's finances by the General Accounting Office.

Describing the report to Congress in 1982, Wilbur Campbell of the GAO said his agency found that EEOC employees untrained in accounting had been given bookkeeping responsibilities. It uncovered cases of employees fighting one another on the job and threats of violence against supervisors. In one instance, guards working for the General Services Administration were called in to maintain order at the EEOC's headquarters.[2]

Some of the EEOC's discrimination cases were at least six years old, and half were over a year old. Thomas's staff later found case files stuffed behind filing cabinets and hidden behind ceiling tiles.

Thomas knew the EEOC was in administrative turmoil when he took over, but he had no idea of the severity. Arthur Fletcher, a black Republican who had worked in the Nixon administration, said he had turned the chairman's job down in Reagan's first year in office after several EEOC employees warned him away because of fierce internal divisions. "They showed me documentation that made it very clear to me that there were saboteurs all over the damn place and you wouldn't be able to trust anybody there," said Fletcher. "It was internal ethnic warfare."

Thomas tasted the agency's internal politics his first day on the job. Shown his new office, he found employees had stripped it of every piece of furniture, save a desk. He was horrified on more than one level. As chairman, he was now responsible for EEOC's condition. "I was thirty-three years old," he recalled. "I'd only been in this town three years, and I see this mess. And it's all going to be blamed on me."

Thomas set out to make his mark at the EEOC in an extremely hostile political environment. Reagan campaigned aggressively against "quotas" and other "preferences" for racial minorities, angering civil rights leaders even before he took office.

After the election Reagan charged the Department of Justice with implementing his campaign pledges on civil rights. EEOC was never even on his radar. Some in his administration did not believe the EEOC should exist, viewing it as excess government regulation. The administration was so slow to fill vacancies on the commission that it operated without a quorum for months before Thomas arrived.

For input on civil rights issues, the White House was more likely to seek out Clarence Pendleton, the chairman of the U.S. Commission on Civil Rights. Thomas, by contrast, was an unknown quantity working in an unappreciated bureaucracy. Privately, Thomas coined his own moniker for his new place of employment: the dungeon.

The administration's early assault on numerical quotas for hiring

minorities and busing to equalize segregated school districts antago-
nized and alarmed civil rights activists. Their anger deepened when
the administration sent out mixed signals about its support for an ex-
tension of the 1965 Voting Rights Act, a capstone of the civil rights
movement.

In the midst of the controversy, the Treasury Department announced in January 1982 that it was overruling a decision denying
tax-exempt status to two private schools with racially discriminatory
policies. One of the schools, Bob Jones University in South Carolina,
banned interracial dating and marriage among its virtually all-white
student body.

The reversal was announced on a Friday, the same day that the Justice Department announced an antitrust settlement requiring AT&T
to divest twenty-two regional subsidiaries. At a meeting the following
Monday, Thomas, then still working for the Department of Education, argued that administration's decision on the Bob Jones case was a
catastrophic mistake. It would be seen as an endorsement of racist
policies and would mark the "death knell" of any attempt to win
blacks' support for administration policies, Thomas argued. A partici-
pant at the meeting blithely told Thomas that Bob Jones was a two-day
story that would blow over in media coverage of the AT&T breakup.[3]

Thomas correctly predicted the backlash. Civil rights leaders,
joined by members of Congress, condemned the decision as
government-supported racism. The ferocity of the reaction was pointedly conveyed to Reagan by Thaddeus Garrett, a domestic-affairs advisor to Vice President Bush. Garrett, who was black, listened to his
pastor over the weekend recount the parable of the snake that bites the
woman who nursed it to health. "Reagan is that snake!" the pastor
thundered.[4] At the urging of Michael Deaver, Reagan's deputy chief
of staff, Garrett told the story to the president in the Oval Office.
"Sir," Garrett said, according to Deaver, "they are calling you the Devil
and a snake." Reagan turned ashen and appeared as though he had
been kneed in the groin.[5]

Even William Bradford Reynolds, the architect of Reagan's civil rights agenda at the Justice Department, came to perceive how badly the administration had blundered in the Bob Jones case. "It was the albatross around our neck for the next two terms," he said. "Everyone could say, 'Well, that's Bob Jones.' And Bob Jones, in two words, was a way to capture fifty words that told everybody we were the biggest racists in the world."

Thomas was doomed even before he began at EEOC. Bob Jones, and all that it connoted, became the filter through which all his actions were scrutinized. He told a friend that he felt as if he had a bull's-eye painted on his back. In setting policy at the EEOC, Thomas was marching headlong into the minefield of the nation's racial politics.

In his first year Thomas all but ignored EEOC policy, turning first to the agency's financial turmoil. "I was scared to death," recalled Willie King. "We figured he'd come in and do like a lot of people and fire everybody." He didn't. One by one, Thomas interviewed everyone in the finance department, King recalled, asking them what they needed to do their jobs. "He didn't do any finger pointing," said King. "He gave us the support we needed."[6] Within months the agency's books were current, and within a year auditors had certified its accounting system as financially sound.

With EEOC's financial picture improving, Thomas focused his attention on the guts of EEOC's work, enforcing antidiscrimination laws in the workplace.

Thomas took an extremely literal interpretation of EEOC enforcement. He believed that employers practicing job discrimination were criminals who should be punished. He was astonished to learn that EEOC's statutory enforcement powers were utterly toothless. Discrimination victims could only be awarded back pay and reinstated to the position they would have held absent discrimination. Thomas routinely fumed that the penalties for stealing someone's job were weaker than those for stealing someone's mail. He told Barry Goldstein, a civil rights lawyer, that EEOC law should be amended to permit jail time for job discriminators—a penalty that not even civil rights advocates were willing to embrace.

Thomas was particularly incensed over a program called "rapid charge," adopted in 1977 as an emergency response to a backlog of more than 130,000 unresolved complaints. The system required EEOC investigators to resolve almost half of all cases through conciliation between the employer and the employee alleging discrimination. The policy solved the backlog problem, but the nitty-gritty details appalled Thomas. Cases were settled with no investigation at all into the underlying charge of discrimination. Many were closed for a hundred dollars and a "neutral" job reference from the employer. Employees who might have been entitled to thousands of dollars in back pay because of discriminatory policies were walking away with a hundred bucks. Worse still, the discriminating employer got off scot-free. Thomas fumed. How could the EEOC call itself an enforcement agency?

Thomas abandoned the rapid-charge system in favor of full investigation of all charges brought to the EEOC. He also shifted money into training for EEOC investigators, most of whom had never learned how to investigate and document their cases.

The change drew money and staff away from the EEOC's systemic programs office, much to the alarm of civil rights advocates. The office investigated mostly large companies for systemic discrimination and generated many of the EEOC's highest-profile cases. Thomas thought the EEOC was focusing too much on those cases at the expense of smaller ones.

He also thought the cases relied too heavily on statistical disparities to prove discrimination. Thomas believed other factors, such as lower educational attainment among minorities, might also explain underrepresentation in an industry. If companies really were discriminating systematically, Thomas believed, private employment attorneys would

be snapping up the cases to sue for damages. Conversely, Thomas believed, no private attorney would take the case of an individual charging discrimination because such cases didn't involve enough money.
Only the EEOC could help that victim.

Thomas's focus on victims also jibed with his belief that government should protect individuals from discrimination, not rectify statistical inequalities in the workplace. This position became a major
bone of contention with civil rights advocates. Thomas's predecessors
forced discriminating employers to pledge hiring goals to pay for discriminatory practices. For civil rights advocates, so-called goals and
timetables had moved tens of thousands of minority workers into the
workforce, giving them job opportunities they might not otherwise
have had.

Ironically, Thomas's first major battles were with members of his own
administration.

In January 1983 the Justice Department challenged a settlement
between the New Orleans Police Department and black officers who
argued that they had been denied promotions. Justice lawyers argued
that the agreement was unconstitutional because all black officers
would be promoted, regardless of whether they had proven discrimination.

Thomas was alarmed. The EEOC routinely negotiated settlements
that included such promotion or hiring goals, often before findings of
discrimination. Justice's position might invalidate scores of EEOC
discrimination agreements. Thomas fired off an angry letter to Attorney General William French Smith, calling the Justice Department's
action "deplorable" and a "serious breach of protocol."[7] With
Thomas leading the charge, the EEOC readied an opposing legal
brief. Justice flatly responded that EEOC had no authority to contra-
dict it in cases involving public-sector employees. Ignoring the memo,

Thomas proceeded with plans to file the brief, knowing that the action would place EEOC directly at odds with the administration.

In subsequent meetings at the White House and Justice Department, Thomas was ordered not to file the opposing brief. The thirty-four-year-old chairman backed down, losing the first of many policy battles with Justice.

In the midst of the controversy, Thomas returned to Georgia to visit his grandmother, who was recovering from an illness in the hospital. His separation from Kathy made the visit especially difficult. Myers and Tena adored Kathy, and as Catholics they objected to divorce. Thomas's Republican politics had also strained his relationship with his grandfather. "That's not the way I raised you, boy," Anderson told Thomas more than once. When Thomas spoke at the Savannah Chamber of Commerce after he joined the Reagan administration, Anderson refused to go, according to his friend Sam Williams. "I don't want nothing to do with him," he told Williams, according to the latter's account.

Thomas explained to his grandfather that he was simply practicing what Anderson had preached about self-reliance, hard work, and independence. At the hospital Thomas complained about the criticism he faced in Washington, pouring out his hurts to his grandfather. "Son, you have to stand up for what you believe in," Anderson said as the two men parted.[8]

At seventy-three Anderson was now an old man, but he did not know how to slow down. He continued to rise before dawn, even though he had retired. He worked on the farm in Liberty County, even in the summer heat. Curators at Fort Morris, a historic site just a mile from the farm, offered him a job that would pay several hundred dollars a month. All Anderson had to do was drive to the fort in the morning to open the gate and return in the evening to lock it. Anderson refused. He'd never worked by another man's clock, and he wasn't going to start now, he told his grandson.

Doctors prescribed Anderson Valium to help disengage the motors that seemed to be constantly running in his mind. His daughter noticed that he seemed to be drinking more of the corn liquor that he liked so much. Leola, now a nurse's aide, took Anderson's blood pressure and was alarmed when it registered off the charts. She urged her father to see a doctor immediately. Anderson snorted, telling her to stay out of his business.

On March 30, 1983, Myers Anderson rose early and spent the morning out in his fields planting, just as he always did in the spring. After lunch he began to feel sick and lay down. He called his cousin Eugene Osborne, who lived up the road. "I don't feel too good," he told Osborne.

Osborne sped over in his car, loaded Anderson inside, and drove him to a hospital in Savannah. Halfway there, Anderson slumped over and stopped talking. By the time Osborne reached the hospital, Anderson was dead. He'd suffered a massive stroke.

Thomas returned to Savannah for his grandfather's funeral in shock, unable to accept that the man who had stood so tall in his life was gone.

"He was a man who didn't rust out, he wore out," recalled Prince Jackson, who delivered Anderson's eulogy, years later. "There are two ways to die. You can sit back and rust like a car on the lawn, or you can run and haul and do good things until you wear out. Myers was one who died because he wore out."

Thomas was still grieving over his grandfather's death when his grandmother died exactly a month later. He cried uncontrollably at her funeral. The loss of both his grandparents, following so closely the collapse of his marriage, pushed him to despair. He felt totally alone.

In his mourning, he was also filled with regret. He condemned himself for once looking down on his grandparents for their lack of formal education. In truth, he now realized, he would be nowhere without their guidance and support. Anderson had pushed him so hard in school. Tena had made sure he got there, with clean clothes and a full stomach. He thought about the many hours he'd spent around the kitchen table with them, doing his homework and listening to the radio.

Thomas began attending church regularly again, after a long lapse from his Catholic faith. Although he did not rejoin the church, he attended morning mass almost daily before work, using the quiet time to collect himself before the long days at EEOC. Over the years, Thomas would return again and again to the last words Anderson ever spoke to him: "You have to stand up for what you believe in."

That spring he flew to Tulsa, Oklahoma, to give a speech at Oral Roberts University. He invited Anita Hill to join him, in part to give her a chance to visit her family on the government's nickel. The law school dean, Charles Kothe, asked Thomas what he would think about Hill leaving his staff to teach at the school. At the time Kothe was concerned about the shortage of black teachers on the faculty. With Thomas's blessing, Kothe asked Hill to apply for a teaching job. Less than three months later she left Washington and the EEOC to take up a career in academics.

Thomas, meanwhile, moved out of Washington that fall and into an apartment in Hyattsville, Maryland, far outside the District of Columbia. He made the move for ten-year-old Jamal, who was now unofficially in his father's full-time custody. He wanted a safer and cheaper place to live. Thomas was spending more than one quarter of his salary to keep Jamal in private school. He had sold his first sports car to defray expenses, relying at least in part on public transportation to get around. Thomas and Kathy were now close to a divorce settlement—it would be final the following summer. Their parting was amicable, and the two remained on cordial terms afterward. By mutual agreement, Thomas obtained custody of Jamal, who continued to spend some weekends with his mother. The divorce nonetheless filled Thomas with anguish. He'd failed as a husband—just like his father—and become part of the very statistic that he decried so strongly in black American life: a broken family.

Whatever his personal failings as a spouse, however, Thomas was determined to do right by his son. He resolved that he would not leave Jamal as his father had left him. He nicknamed Jamal "the Mighty Dude" and tried to build up his self-confidence. Parenting became an obsession, and he was constantly seeking tips from other parents. Thomas was especially keen to know how he could compensate for the absence of a full-time mother in Jamal's life. He often compared notes with Harry Singleton, another divorced father, according to Nancy Altman, a friend. On Mother's Day Thomas and Singleton exchanged Mother's Day cards, she recalled.

In his first two years at the EEOC Thomas mostly avoided the limelight, despite White House complaints that he wasn't doing enough to defend the administration. "They wanted Thomas's face and his standing as head of the EEOC in their storehouse of ammunition," said reporter Juan Williams, who now covered the White House for *The Washington Post*. "Thomas was very discerning and would pick his arguments."

Williams recounted one encounter between Thomas and William Bradford Reynolds in the White House mess. Reynolds was razzing Thomas, urging him to be more aggressive in supporting the administration. "Don't tell me what to do, Brad," Thomas snapped. "All I have to do is die and stay black."[9]

The administration's policies toward racial minorities, particularly its opposition to affirmative action, discomfited Thomas. Despite his own reservations about affirmative action, he bristled at the way the administration was framing the debate. He believed that attacks against affirmative action were attacks against black Americans. The administration's arguments also angered him because they seemed to ignore three hundred years of history. Discrimination against blacks had been written into laws, sanctioned in the workplace, and burned into the American psyche. For the administration to proclaim that the nation no longer needed preferential policies toward minorities was outrageous, Thomas thought. Worse still, ending affirmative action was the sum total of the administration's agenda. No one talked constructively about solutions to the lingering effects of past discrimination.

At first Thomas tried to stake out a middle ground, lacing his public speeches with criticisms of both the civil rights community and the administration. He bluntly scored the Reagan officials for using terms like "reverse discrimination" and "racial spoils system" to pander to "an undercurrent of racism that we know still exists in this country."[10] He told the civil rights community that its emphasis on busing and affirmative action distracted the nation from bigger problems facing black Americans, such as poor-quality schools.

In loftier moments Thomas attempted to fashion a historical narrative that explained how Americans became so deeply divided over the legacy of discrimination. The civil rights movement prevailed, Thomas believed, because it resurrected the fundamental notion of fairness and equality embodied in the American ideal. Yet the national consensus on equality, so passionately articulated by Martin Luther King, Jr., in his 1963 "I Have a Dream" speech, unraveled in the early 1980s over the remedies to past discrimination, Thomas thought. He argued that the country had moved from demanding fair play for black Americans to insisting on preferential treatment for them. White backlash over special breaks for minorities dissolved the national consensus of the King era.

Yet he was still ambivalent about doing away with racial-preference programs altogether. "Is it right to be color-blind in a nation that has seen color for so long?" he asked in a 1983 speech. "Is it right to eliminate preferences in the civil rights area when preferences is the name of the game in so many other areas? Is it right to call for fair play when fair play has never existed?"[11]

He would wrestle with the question for the next twenty years, but already he was inclining toward the color-blind legal philosophy that he now promotes on the Supreme Court.

"I opt that we try to adhere to race-neutrality and fairness," he said in another 1983 speech. "That is, treating individuals as individuals, neither preferring nor deterring on the basis of group characteristics. But I choose this option knowing full well that fairness, though an underpinning of this country, has never been a reality. I choose the option with the painful awareness of the social and economic ravages that have befallen my race and all who suffer discrimination. I choose this option with the enduring hope that this country will live up to the principles enunciated in the Bill of Rights and the Constitution of the United States: that all men are created equal, and that they will be treated equally."[12]

Thomas continued to support affirmative action remedies in EEOC settlements, but he often found the details troubling. Black workers were getting job opportunities through affirmative action, but often at others' expense. Should a black worker get a promotion over a white employee with the same qualifications? If numerical goals were set for hiring and promoting minorities, who decided what was the "right" number? Should the goal include American Indians, Hispanics, and other victims of discrimination? The fairness issue was so difficult, Thomas said in a speech at the University of Virginia in October 1983, "that even King Solomon would be frustrated."[13]

Thomas also thought the emphasis on affirmative action diverted attention from other problems. Studies showing that up to 40 percent of black youth were functionally illiterate alarmed him. Other data showed that one in six minority teenagers was not even in school. Half of black teenage girls were projected to be mothers before the age of twenty. What good was affirmative action if black Americans weren't even finishing high school? Thomas wondered. What good was it if 42 percent of black college students flunked out?

"What we are not dealing with is the fact that the school systems still aren't educating our kids," Thomas told *The Crisis*, the NAACP's magazine, in 1983. "Now, you can say it's racism. You can say all sorts of things. I don't care. Kids are still not getting an education."

Thomas's ruminations about education, teen pregnancy, and black-on-black crime put him at odds with civil rights leaders, who in his perception were more preoccupied with political rights. Thomas also objected to the way in which the civil rights establishment blamed all problems of black Americans on racism; he believed that practice convinced black young people that success was hopeless.

"We need to begin to think about what our history meant to us and how we made it for four hundred years and suddenly, fifteen years now, we can't make it," Thomas continued in *The Crisis*. "What happened? Who let the wind out of our sail?

"Maybe there are no more lunch counters out there to desegregate, but there's a whole life out there to desegregate. And if we give up now and keep spending our energies blaming somebody else, are we ever going to make it? Sure, racism is a part of it. Sure discrimination is a part of the problem. But people are just feeding us this stuff that's hopeless, making the kids say, 'Why should I try?' "[14]

On October 18, 1983, General Motors settled a ten-year-old discrimination case with the EEOC and agreed to pay $42.5 million in back pay and other commitments. The record settlement included goals and timetables for hiring and promoting blacks, women, and Hispanics. GM would also provide $15 million in educational scholarships, two thirds of which would be directed toward historically black colleges. The scholarship provision reflected one of the "alternative" approaches to affirmative action that Thomas believed might work. By directing the scholarships toward historically black colleges, Thomas was ensuring that they would benefit black students. Some civil rights and women's organizations, however, charged that the back-pay provisions were inadequate. "It's a travesty," fumed Nancy Kreiter, director of Women Employed, an advocacy group that kept close tabs on the EEOC.[15]

After eighteen months on the job, Thomas saw his relationships with civil rights group take a sharp turn for the worse. Part of it was his own fault. From the beginning, he made little effort at dialogue with the advocacy groups. Cathie Shattuck recalled that he rarely scheduled meetings with civil rights advocates to discuss their concerns. In those meetings that civil rights leaders insisted on, he appeared reserved and uncomfortable, recalled William Robinson, then director of the Lawyers' Committee for Civil Rights Under Law. "It was always a question whether you were going to break the ice with him," said Robinson.

In fact, Thomas was uncomfortable around civil rights advocates. They represented what he had chosen not to become, at least in the way that they had chosen to define the role. To him they embodied an aggressive, confrontational approach that he simply couldn't adopt. Leaders of the movement changed society by doing the one thing Thomas was told he should never do: complain. As much as Thomas embraced the ends of the civil rights movement, he was never comfortable with the means.

Thomas expressed his discomfort in a widely quoted news article from 1984. Civil rights leaders simply wanted to "bitch, bitch, bitch, moan and moan, whine and whine," Thomas quipped.[16] Civil rights leaders were not endeared.

Thomas also believed that the EEOC had been too cozy with the civil rights establishment, and he was determined that the agency remain neutral. He believed that coziness drastically undercut the agency's ability to deal forcefully with workplace discrimination. For their part, civil rights advocates quickly realized that Thomas mostly ignored their input.

By 1984, advocacy groups were also troubled by EEOC case data. Cases dismissed for "no cause" were rising sharply, while the number of those settled for cause was declining. The number of class-action suits filed under Thomas had fallen off dramatically. The changes were a direct result of Thomas's new enforcement policy. The number of cases without merit was increasing because EEOC investigators were actually investigating cases, which hadn't been done under the rapid-charge system. The emphasis on investigation also meant the cases took longer, so there were fewer settlements.

Yet Thomas was accomplishing what he set out to do: compensate actual victims of discrimination. Average settlement awards nearly tripled. The decline in class-action lawsuits also reflected Thomas's philosophical problems with those types of cases.[17]

Thomas's persistent criticisms of goals and timetables also irked civil rights advocates. A 1979 Supreme Court decision had clearly spelled out that goals and timetables were legal remedies. Civil rights advocates wanted the EEOC to use goals and timetables in all big settlements. Thomas believed they should be used in only some cases.

While the advocates had little sway with Thomas, they had many friends on Capitol Hill, particularly in the Democratic-controlled House of Representatives. At the urging of civil rights organizations, several House committees began an aggressive review of the EEOC's enforcement policies. In his first two years at the EEOC, Thomas testified before Congress more than twenty times, a number that would climb to more than fifty by the time he left in early 1990. Thomas hated the experience. He became morose and sullen on days that he was hauled in to testify, referring to the sessions as his trips to the "woodshed." He believed advocacy groups were using Congress as a cudgel, forcing him to make changes that they couldn't make him do on their own.

Thomas invariably put on his game face before Congress, sitting stiffly at the witness table and refusing to concede an inch. Committee members badgered him with questions; Thomas badgered back. He hated being bullied.

His aggressive stance, however, angered Democrats and set the tone for how he was viewed later on as a nominee to the Supreme Court. "He would just stomp over and be outrageous," said Patricia Schroeder, the former congresswoman from Colorado. "He just relished going around and poking people in the eyes who he didn't agree with."

Thomas also used the hearings for his own political aims. Hearings gave him a public forum in which to distance himself from EEOC policies that were opposed by the Reagan administration. The EEOC, for example, repeatedly approved settlements that included hiring goals and timetables. Yet he repeatedly told members of Congress that he didn't believe in goals and timetables. The message was clear: Whatever EEOC's organizational policies, Thomas was still on the GOP reservation.

Thomas's toughness in committee hearings also won him powerful Republican allies, including Utah senator Orrin Hatch, chairman of the Senate Labor Committee. Hatch liked Thomas's guts, a point that was driven home when Thomas wagged an angry finger at Massachusetts senator Edward M. Kennedy. Kennedy had told Thomas he should be ashamed of his handling of employment discrimination. Thomas cut Kennedy off to say his grandfather hung only three pictures at home—Martin Luther King, Jr., Jesus Christ, and President John F. Kennedy. It was President Kennedy who would be ashamed of his brother, Thomas angrily retorted. "Kennedy was dumbfounded," Hatch recalled later.

Thomas also angered women's organizations by refusing to endorse the concept of comparable worth. The Equal Pay Act, which the EEOC enforced, required that men and women receive equal compensation for equal work. In the early 1980s women's advocates aggressively promoted the idea that certain jobs traditionally held by women were of comparable value to different jobs held by men. The Reagan administration, backed by potent business lobbies, opposed comparable worth. So did Thomas. "He really wasn't interested in dialogue about it," said Schroeder, whose House committee repeatedly urged Thomas to embrace comparable worth. "We just gave up, because we decided all he wanted was theater. He thought he was getting points by having the balls to come over and spit in our face, and that's what he did."

Thomas almost seemed to mold his public persona on his favorite fictional character, Howard Roark, the misunderstood genius in Ayn Rand's novel *The Fountainhead*. Thomas frequently screened the movie based on Rand's book for EEOC employees over lunch. The film is a didactic, almost corny rendering of Rand's objectivist philosophy of enlightened self-interest.

In the movie Gary Cooper plays an architect who chisels granite in a rock quarry rather than design buildings that compromise his own principles of architecture. "My reward, my purpose, my life, is the work itself," Roark declares early in the film. "My work done my way! Nothing else matters to me." When one of his buildings is changed without his consent, Roark dynamites it. He stands trial for the crime and, acting as his own lawyer, delivers a passionate closing argument to the jury. "I came here to say that I do not recognize anyone's right to one minute of my life, nor to any part of my energy nor to any achievement of mine," he says, pacing the courtroom. "The world is perishing from an orgy of self-sacrificing. I came here to be heard, in the name of any man of independence still left in the world. I wanted to state my terms: I do not care to live or work under any others. My terms are a man's right to exist for his own sake!"[18]

Some days, Thomas wasn't Clarence Thomas at all. He was Howard Roark, the man who built buildings according to no blueprint but his own.

The prickly iconoclast contrasted sharply with the everyman reputation that Thomas earned at EEOC headquarters. He regularly cruised the hallways to chat with employees, trading one-liners or inquiring about family. As EEOC chairman, Thomas had a government car and driver, but he refused to sit in the backseat, instead joining the driver up front. "He was very down-to-earth," recalled Joan Ehrlich, one of Thomas's district directors. "He wanted people to be innovative. He wanted people to be assertive."

Thomas's staff meetings every morning were optional, but Linda Rule, a former special assistant, said staff almost always attended. Thomas regaled them with endless stories. "He was like a lightbulb in the center of any room," said Rule. "He didn't have an adoring staff because he gave us perks. We were just very taken with his uniqueness."

Thomas frequently stopped by Rule's desk at the end of the day to stuff his pockets with candy from a bowl she kept on her desk. Rule, who was Jewish, bonded with Thomas over their mutual experiences of being excluded or discriminated against. Thomas was especially sympathetic to Rule's first job interview out of school, when she was turned down point-blank because of her faith. He also took an interest in Rule's daughter, who was born with a growth problem. Rule recalled, "He kept saying to me, 'Reward her. Let her know that she's special and you know it.'"

Thomas also demanded excellence from staff and didn't easily tolerate mistakes. "He set a very high standard for himself and everyone around him," said Rule. "We all had a fear of disappointing him, so we all tried very, very hard."

In commission meetings Thomas explained his policy positions but never twisted arms or brokered deals to get his way. "Before he would make a decision he would try to get as much input from different sides," said Leonora Guarraia, one of his senior staff members. "He wanted the policies to be well balanced, as opposed to going off in any one particular direction." The approach won over skeptical employees, Guarraia said. "Even people who didn't agree with him politically had the highest respect for him."

Reagan's landslide reelection over Democrat Walter Mondale meant Thomas wasn't out of a job. Yet by 1984, he remained relatively unknown to the president, and he felt other key members of the administration still regarded him with suspicion. He was rarely, if ever, consulted on major policy matters, even those affecting minorities. Unsolicited advice, such as his counsel on Bob Jones University, was ignored. At White House functions Thomas was routinely seated at tables near the back of the room, according to Armstrong Williams, his former aide. Indeed, Ricky Silberman, the EEOC's newest board member, appointed by Reagan after his reelection, had the vague sense that she was being offered the job of EEOC vice chairman for an unstated reason: to keep an eye on the chairman. There was even talk at the White House of forcing Thomas's resignation, according to *Washington Post* reporter Juan Williams.

Thomas felt the disrespect acutely. "EEOC is the place where they send you when they really just want to give you a job, but they don't expect that much from you," he told Armstrong Williams. "Think of anybody who has ever left the EEOC who is revered and accomplished and sought after—nobody."

But Thomas repeated something else to Williams: "You and I are going to change that," he said. "We're going to become its greatest alumni. We'll show them. We'll show them."

Patience, Thomas told Williams, was an honest man's revenge.

CHAPTER 11

FAST TRACK

While Thomas toiled in relative obscurity at the EEOC, a major fault line had opened between Republicans and Democrats over the federal judiciary. The political divide held serious repercussions for Thomas's future.

In his first term Reagan vowed to appoint judges who would exercise "judicial restraint." Under former chief justice Earl Warren, Reagan and other conservatives believed, the Supreme Court had upset the constitutional separation of powers by "making" law properly left to Congress or the states. Conservatives believed that the Court's rulings, particularly involving individual rights, were often based more on the policy beliefs of the justices than on the language of the Constitution. *Roe v. Wade*, the 1973 decision legalizing abortion, provoked special scorn. The decision's basis—a women's right to privacy—wasn't explicitly mentioned in the Constitution.

Roe became a key litmus test used to pick judges supportive of judicial restraint, said William Bradford Reynolds, who screened nominees at Justice. "We were never of the view that we should get judges to overturn *Roe v. Wade*. But what we did do was we said, 'Tell us what you think of that decision and why.'" Nominees embraced the *Roe* decision at their peril. "Everybody who came in would always say, 'I'm a strict constructionist,'" said Reynolds. "And you can ask people different legal questions and . . . find out pretty quickly whether they are or if they are just blowing smoke."

The Reagan judicial standard alarmed liberals, who viewed strict construction of the Constitution as a threat to individual liberties, particularly abortion rights. Liberals argued that the Constitution's framers drafted a simple document that could grow with the nation, and judges needed to read the document flexibly. While the Constitution made no explicit reference to a right to privacy, various parts implied that that right existed.

Key constituencies in both parties provided the raw emotion that animated the legal debates. Religious conservatives such as the Reverend Jerry Falwell fueled the Republican Party's opposition to abortion and the drive to appoint judicial conservatives. "We will work for the appointment of judges at all levels of the judiciary who respect traditional family values and the sanctity of human life," the party's 1980 platform said.[1] On the other side, women's groups drove the Democrats equally hard to take a firm stand against anti-abortion judicial nominees.

Robert H. Bork, one of Reagan's first major judicial selections, exemplified Reagan's approach to the judiciary. Reagan named Bork to the U.S. Court of Appeals for the District of Columbia Circuit in 1981. A decade earlier Bork had essentially drafted the conservative blueprint for judicial decision making in his *Indiana Law Review* article "Neutral Principles and Some First Amendment Problems." The article, written two years before *Roe*, attacked a 1965 Connecticut case that became the foundation for Justice Harry Blackmun's landmark *Roe* opinion.

. Many of Reagan's early appointments received little opposition in the Senate, which Republicans controlled for Reagan's first term. But his reelection in 1984 changed the terms of the debate. Democrats and liberal interest groups realized the sweeping potential Reagan now had to remake the federal judiciary. The Alliance for Justice, a liberal advocacy group, launched its Judicial Selection Project to research

Reagan appointees and generate opposition to those who were anti-choice and conservative on social issues. The group was to play a major role in the campaign to defeat Clarence Thomas seven years later.

In his 1985 book *God Save This Honorable Court*, a liberal Harvard law professor named Laurence Tribe fired another warning shot on the Senate's "advise and consent" power to confirm Supreme Court justices. Tribe argued that the Senate had largely abdicated its consent role, giving the president too much leeway in molding the judiciary. For Tribe, the Senate's basic criteria for evaluating judicial nominees—competence and moral character—were too narrow. He argued that a nominee's ideology was also fair game and that nominees should be quizzed on *Roe*, sex discrimination, and affirmative action, along with freedom of speech. Tribe's arguments fundamentally altered the way Democrats and liberals viewed the judicial selection process, and he became an advisor to Senate Democrats in Supreme Court confirmations.

On the other side of the aisle, Edwin Meese, Reagan's second-term attorney general, expounded the administration's judicial philosophy. "It has been and will continue to be the policy of this administration to press for a Jurisprudence of Original Intention," Meese said in a widely quoted 1985 speech. "We will endeavor to resurrect the original meaning of constitutional provisions and statutes as the only reliable guide for judgment."[2] In the speech, Meese urged federal judges to closely examine laws that encroached on the powers of state governments, as well as to be skeptical of the rights of criminal defendants and the barriers between church and state not anchored in the Constitution.

The jurisprudence of original intention, as Meese called it, coincided closely, though not entirely, with Thomas's own constitutional thinking.

As a conservative, Thomas was keenly tuned to the debate over the judiciary, and he was surrounded by people telling him that he was destined for the federal bench. Jay Parker says he told Thomas he was headed for the Supreme Court from the moment he landed his first administration job—provided Thomas played his cards right and stayed out of trouble. Reynolds said Thomas was always "in the mix" for a possible judicial appointment because of his conservative credentials, race, and age. Reagan officials especially wanted young appointees. "The younger the better," Reynolds said. Armstrong Williams, Thomas's aide, said he used to tell Thomas repeatedly that he was being groomed for the Supreme Court. Thomas's response, Williams said, was "Never."

"They will never appoint me to the Supreme Court," Thomas said. "I'm too conservative. Even for these Republicans."

Reagan's second term marked a sharp escalation in Thomas's disagreements with Congress, and he was increasingly on the defensive. In late 1984 and early 1985 he stirred up a hornets' nest of controversy over plans to revise EEOC's uniform guidelines on employee-selection procedures, the rules the agency followed to litigate cases.

Seemingly arcane, the uniform guidelines were in fact the legal codification of twenty years of antidiscrimination law. Their application had forced employers to eliminate or modify promotion tests that kept out minorities and women, opening up tens of thousands of jobs to discrimination victims. The NAACP Legal Defense Fund and the Lawyers' Committee for Civil Rights Under Law had devoted literally millions of dollars and thousands of hours to litigating the cases that comprised the guidelines. In seeking their revision, Thomas was declaring war on the nation's most powerful civil rights advocates.

Congressional Democrats hauled Thomas in for a fresh round of hearings over the proposed revisions. Thomas faced a blizzard of questions from hostile Democrats when he appeared before a congressional committee in October 1985. Representative Matthew G. Martinez, Democrat of California, accused Thomas of seeking to revise the guidelines in secret, without public hearings. Thomas went ballistic. "I don't have the slightest idea what you are talking about," he said, fuming. "I don't have the slightest idea what the rumor mill has or the rhetoric mill has or what ghosts and goblins have created out there."

Later, glaring at Martinez, Thomas barked, "It often seems like I am on trial here!"[3]

Thomas eventually retreated from revising the guidelines, concluding that the effort was simply too controversial, according to Philip Lyons, a former special assistant to Thomas.

No longer a neutral bystander trying to pick his way across Washington's treacherous racial minefield, Thomas increasingly perceived himself a hardened veteran of the war. Being part of an administration largely despised by blacks was "my burden to bear and my cross to carry," Thomas told a mostly black audience in Compton, California.

Thomas chided blacks for having children too young, not taking care of them, and shirking responsibilities. "We have become more interested in designer jeans and break dancing than we are in obligations and responsibilities," he said. Government policies, Thomas said, had turned black people into beggars and charity cases. Rather than seizing opportunity, he said, black Americans were hiding behind excuses.

"Not only do you have to contend with the ever-present bigotry," he continued, "you must do so with a recent tradition that almost requires you to wallow in excuses. You now have a popular national rhetoric which says that you can't learn because of racism, you can't raise the babies you make because of racism, you can't get up in the morning because of racism. You commit crimes because of racism. Unlike me you must not only overcome the repressiveness of racism, you must also overcome the lure of excuses. You have twice the job I had."[4]

Thomas also hadn't abandoned his separatist streak. "I am part of the grand experiment, having started by competing against the odds and succeeding, then having my personal achievements inextricably woven in the whole cloth of racially preferential social policies," he said in one 1986 speech. "I have known segregation, racism, and discrimination. I have survived the rejection of segregation and the trauma of integration. I have been both deterred and preferred by racially conscious policies. I have been the guinea pig for many social experiments on minorities.

"To all who would continue these experiments, I say, 'Please, no more. Please leave me alone . . .' "

Thomas conceded that affirmative action resulted in better opportunities for minorities. But now, he argued, the negative assumptions underlying affirmative action were more pernicious than the economic benefits.

It "is assumed that, for whatever reasons, we cannot compete with whites when it comes to intellectual pursuits," Thomas said. "My friends, I will not concede my intellectual inferiority or my son's for socioeconomic gain. We are not objects of charity and refuse to be treated as objects of disdain. Though denied equal opportunity, we are inherently equal. I cannot trade what God has given for what man has promised. Equal opportunity was taken away based on pernicious, racist assumptions of inferiority, and I categorically reject any promises of advantage based on the selfsame obnoxious, though now well-intentioned, assumptions. For me, this concession has been, and will always be, too great a price to pay."[5]

In mid-1986 Thomas was confirmed for a second four-year term as head of the EEOC. Walking him over to the Senate hearing room for his confirmation, Danforth asked his protégé why he wanted to stay on. "The job's not done," Thomas told Danforth simply. Yet his stubbornness also factored in. Leaving would make it appear that he was caving in to his critics. "Winston Churchill was asked, 'Why did you become prime minister?' " Thomas said in a 1987 interview. "He said, 'Ambition.' Well, why did you stay so long? He said, 'Anger.' That's one of the reasons I went back up for reconfirmation. You're not going to run me out of town. I'm going to stay right here."[6]

Reappointment gave Thomas an opportunity to set new priorities, and he began with his own self-improvement.

At the age of thirty-eight, Thomas held political beliefs that were shaped almost exclusively by his own life experiences and the lessons he gleaned from his grandparents' struggles with Jim Crow. He believed in self-reliance, personal responsibility, and individual freedom. Yet he was also painfully aware that his worldview lacked breadth and depth. His academic studies were focused in only two areas: the law and English literature. Indeed, Thomas's political philosophy was arguably more influenced by literature than by history or political science.

After his reappointment, Thomas put two conservative political scientists on the government payroll to help him flesh out his thinking. Ostensibly, Ken Masugi and John Marini were hired to write speeches for him, and they did. But their real function was to serve as his intellectual mentors, acting almost like graduate-school teaching assistants leading a philosophy seminar. They assigned Thomas readings in history and political science. Thomas frequently grabbed either Masugi or Marini for lunch breaks, or asked them to accompany him whenever he had to drive somewhere in town. He usually took a midafternoon cigar break outside the EEOC office. Masugi or Marini joined him for long political discussions on the plaza outside the building. "He was tremendously inquisitive and extremely diligent," recalled Marini.

Both Masugi and Marini were intellectual disciples of Leo Strauss, the German philosopher who decried the rise of moral relativism and inspired a generation of conservative political theorists. Masugi studied under Harry Jaffa, one of Strauss's foremost American champions, and devoted considerable scholarship to illuminating Alexis de Tocqueville's famous *Democracy in America*. Marini's specialty was the constitutional separation of powers. Shortly after he began working for Thomas, he edited a collection of essays on the "imperial Congress," which argued that liberal interest groups and power-hungry lawmakers had toppled the balance of power envisioned by the Constitution's framers. Both men were strong proponents of limited, constitutional government.

Marini and Masugi put Thomas on a diet of European and American thinkers. They assigned him Tocqueville's *Democracy in America, The Federalist Papers,* and treatises by John Locke, James Madison, and Thomas Jefferson. As a member of the executive branch, Thomas was particularly interested in defining the limits of administrative power. He challenged his tutors with lofty questions about the nature of government in a democratic society. When was administrative rule making a legitimate exercise of executive power, and when did it usurp the power of the people's elected representatives? What were the limits of freedom? How much freedom could government regulate and control? Thomas reveled in the discussions and acted like a man who had suddenly been given the gift of doing his formal education all over again. "He was a great intellectual," Masugi recalled. "He would never call himself a great intellectual, but he certainly had that graduate student's spirit."

The influence of his new tutors was evident almost immediately. Thomas's speeches suddenly were filled with references to great philosophers and esoteric musings about freedom and democracy.

Slavery and its legacy of segregation consumed much of his thinking yet presented Thomas with an almost unsolvable dilemma. How could he believe in a government that once upheld the bondage of human beings? Institutionalized racism in America made a mockery of all that America stood for, he believed. His allegiance to conservative orthodoxy made the questions all the more difficult, particularly regarding the Constitution.

Conservative orthodoxy in the judiciary, as Meese had so forcefully articulated, commanded adherents to hew closely to the Constitution's literal meaning at the time it was adopted. For Thomas, it seemed impossible to reconcile original intent with history. The Constitution had in fact condoned slavery. The rights enumerated in the Bill of

Rights were written by and for white, property-holding men. The post–Civil War amendments outlawing slavery had been interpreted to permit segregation. To interpret the Constitution literally, to rely on the intentions of the framers was to accept the notion that black Americans were never intended to have equal rights and freedoms.

Thomas found the answer to his dilemma in the Declaration of Independence. He came to view the document, with its pronouncement that all men were created equal, as the ultimate expression of the nation's founding principles. The approach took him deep into an arcane, yet controversial, set of legal theories known as natural law, the notion that a higher law, God's law, was the ultimate standard against which man's laws were to be judged. Thomas believed that the Declaration expressed the higher-law concepts of the framers.

Thomas's thinking about natural law was heavily influenced by Strauss disciple Harry Jaffa, a German-born philosopher who taught at the University of Chicago, and Jaffa's research on President Abraham Lincoln.

Natural law, which later became so controversial in Thomas's nomination to the Supreme Court, was an easy fit for Thomas. The belief in a higher law was what sustained his family against the moral contradiction of segregation. God's law taught Thomas to believe in his innate equality, even when the white man's law said he was inferior. Segregation was wrong, Thomas grew up believing, because it offended God's law. He found support for his beliefs in the writings of Dr. Martin Luther King, Jr., and was heavily influenced by King's famous "Letter from a Birmingham Jail," in which he argued that God sanctioned opposition to unjust laws.

In Lincoln Thomas found a passionate spokesman for the ideal of American freedom and equality. In King he found a way to expunge the blot of slavery and segregation while holding fast to a belief in the rule of law. All that was left was to find a constitutional principle that tied the two together.

For that, Thomas turned to a nineteenth-century Supreme Court justice named John Harlan, the lone dissenter in the Court's historic *Plessy* v. *Ferguson* case, which effectively legalized segregation and all the punitive Jim Crow statutes that followed. Homer Plessy challenged a Louisiana law requiring separate train cars for black and white passengers. The Court ruled against the black plaintiff, saying that separate but equal facilities were no violation of the Constitution.

Justice Harlan wrote a dissenting opinion in *Plessy*, arguing that segregation was unlawful because the Constitution created no classes or racial distinctions among Americans. "Our Constitution is color-blind, and neither knows nor tolerates classes among citizens," he wrote.

Thomas chose to ignore elements of Harlan's dissent that clearly reflected Harlan's own racism. Indeed, Harlan prefaced his comments about the color-blind constitution with the statement "The white race deems itself to be the dominant race in this country. And so it is in prestige, in achievements, in education, in wealth and in power. So, I doubt not, it will continue to be for all time if it remains true to its great heritage and holds fast to the principles of constitutional liberty."[7] Thomas attributed the view to those of Harlan's era.

What mattered to Thomas was the principle Harlan was articulating: that the Constitution, properly understood, saw no color. The idea squared perfectly with Thomas's own beliefs about the innate equality of black and white Americans. And it solidified his view that what had perverted the notion of equality in America were color-conscious laws that denied black Americans their fundamental rights and freedoms.

Thomas's fulminations about the nation's founding principles also drove him farther to the right in his support of limited government. He became more shrill in his denunciations of welfare and other entitlement programs of the Great Society. He called welfare a "narcotic" that was addicting black people and robbing them of their independence.[8] In April 1987 Thomas all but said that the nation was careening toward socialism, and he belittled affirmative action as nothing more than revenge against white Americans for centuries of exploitive policies toward black Americans. "Retribution isn't justice, no matter how good it may make some feel," he said.[9]

Ever so perceptibly, Thomas's stock was beginning to rise in Washington. Several key contacts helped pave the way. One of the most important was EEOC vice chairwoman Ricky Silberman, a longtime Republican who moved in the upper circles of the Republican Party. Silberman introduced Thomas to a world of GOP contacts previously closed to him. She invited him to dinners at her Georgetown town house, often including Thomas Sowell when he visited Washington. The dinners also included Silberman's husband, Laurence Silberman, whom Reagan named to the D.C. circuit court of appeals in 1985 after a long career in two previous Republican administrations. Along with his wife, Silberman was extremely close to some of Reagan's top legal advisors. Silberman, like Reynolds, believed Thomas might make a good judge.

Thomas also developed a friendship with C. Boyden Gray, the most critical contact in his path to the judiciary. At the time, Gray was counsel to Vice President Bush and himself a rising figure in the Republican Party. Gray, who clerked for Chief Justice Earl Warren in his final year on the bench, liked Thomas immediately and quickly saw him as a potential candidate for the federal judiciary.

Thomas was developing a critical mass of support for a slot on the bench. Increasingly, the question was not whether to nominate Thomas to the judiciary, but when.

CHAPTER 12

VIRGINIA

Thomas was adamant: He was not going. Ricky Silberman had seen the stubbornness before. Thomas glowered, crossed his arms, and dug in his heels. "I'm not going to New York City," he said with characteristic obstinacy. In the spring of 1986, The Anti-Defamation League in New York had organized a roundtable to discuss affirmative action. Jeff Zuckerman, the EEOC's acting general counsel, was supposed to represent the agency, but he had canceled at the last minute.

"I'll get Clarence to go," Silberman had said.

"He'll never go," Zuckerman had responded.

"Leave it to me," Silberman had told him.

"I'm not going," Thomas repeated to Silberman now.

"You're going," she said.

The morning of the conference the two boarded a shuttle to New York. The topic for the forum was "How Long Does America Need Race-Preference Policies?"

Thomas and Silberman took their seats around a large conference table at the ADL offices in Manhattan. It was a diverse group of participants. Bayard Rustin, the aging strategist of the civil rights movement and one of Martin Luther King, Jr.'s confidants, was there, along with author Midge Decter. Casting his eye around the table, Thomas paused to look at a young woman who introduced herself as Virginia Lamp, a labor lawyer for the U.S. Chamber of Commerce.

That's Virginia Lamp? Silberman thought to herself during the introductions. Lamp was frequently mentioned in the press as a spokeswoman for the Chamber. Silberman had pictured a woman in her fifties, an economist perhaps. Here instead was a tall young woman with high cheekbones and dazzling blue eyes. Silberman was struck by her perfect skin—smooth and white, like alabaster.

The conference broke for a working lunch, and Thomas made his way around the table to introduce himself to Lamp. At the end of the day Silberman went looking for a telephone before catching a cab to the airport with Thomas. When she returned to the conference room, Thomas was gone.

At that moment, Thomas was sharing a cab ride to the airport with Lamp. Lamp had read about the EEOC chairman and later remembered asking him how he coped with negative press. He pulled out a prayer from his wallet. He's religious, she thought to herself approvingly. As they got out of the cab, she blurted out, "I've got some black women friends I'd like to introduce you to." Thomas laughed. It was the beginning of an improbable romance.

Virginia Lamp was born and raised in Omaha, Nebraska, the youngest of four children. Her father worked for the U.S. Army Corps of Engineers before breaking off to start his own engineering firm. Her mother stayed at home raising the children. When Virginia was in the seventh grade, the Lamps moved to a house on a lake outside Omaha. Her childhood revolved around the water—swimming, sailing, and water-skiing.

Lamp's mother was active in the Nebraska Republican Party and carried the political genes in the family. Virginia attended county GOP meetings with her mother and was volunteering on local Republican campaigns long before she could vote. In high school she joined student government and was elected an officer of her senior class.

To classmates she was known as "Warrior Woman," after her high school sports teams, the West Side Warriors. At football games Lamp dressed in a cheerleader's skirt, grabbed a sword and shield, and paraded with the school marching band.

Even in high school her ambition was to be a member of Congress, representing Omaha. She enrolled in a small women's college in Virginia, just outside Washington, to be close to the nation's capital, but transferred to the University of Nebraska in Lincoln after a year to be closer to her boyfriend, a pre-med student at Creighton University in Omaha. Then she switched to Creighton, majoring in business communications and political science, before enrolling in Creighton's law school.

In her freshman year Lamp headed the youth campaign of Hal Daub, Jr., a Republican candidate for Congress. One of her favorite professors at Creighton, G. Michael Fenner, was a Democrat who worked against Daub. "She was very thoughtful in her politics," said Fenner, who has remained close to his former student. "She doesn't like you or dislike you because of what you think. She likes or dislikes you for the kind of person you are."

Daub won the 1980 election, and Lamp went to Washington to work for him for eighteen months before returning to Omaha to finish her law degree. She graduated in 1983 and took a job as Daub's legislative director. A year later she went to work for the U.S. Chamber of Commerce. Two years after that she met Clarence Thomas.

Before their first date Lamp resolved to tell Thomas that they should just be friends. He took her to see the movie *Short Circuit*, a comedy about a military robot that becomes a pacifist after being hit by lightning. Watching the film, Lamp heard Thomas laugh for the first time. She'd never heard anyone laugh so spontaneously and with so much gusto. Soon she was laughing at his laughing.

At dinner she told him that she had a boyfriend and that they should just be friends. He agreed, telling her that he was also dating someone. Without the pressure of romance, the two talked long into the evening. By the end of the dinner, Lamp was in love. She took the subway home that night and called the man she was dating, breaking up with him over the telephone. Thomas called her for another date. He picked her up in his black Camaro, son Jamal in tow. The three sped off to Baltimore, where they walked around the city's inner harbor. Later, just before he kissed her for the first time, Lamp remembered Thomas apologizing for what he was about to do, for drawing her into the "race thing," as he called it. Forget about it, she told him.

Lamp had lived most of her life far removed from the nation's racial battlegrounds. Growing up, she had all white neighbors. All her high school classmates were white. Race was discussed only rarely in her household. Her parents appeared to be neither hostile to blacks nor enthusiastic about them. Black people were simply not a part of their lives.

Lamp's first realization that race mattered to her family occurred after she went out to dinner with a black law-school classmate in Omaha. Her mother was furious.

When she fell in love with Clarence Thomas, Lamp realized she might confront similar antagonism. Calmly, she sat down and wrote her parents a long letter. She explained that she had met the man of her dreams, someone who loved her and took care of her better than anyone she had ever known. She hoped they could accept him. If they could not, she said simply, they would lose a daughter.

Over lunch one day Thomas told Ricky Silberman that he was engaged. She assumed Thomas had proposed to the woman he had been dating before Lamp. "He said, 'You're not going to believe this. It's Virginia Lamp,'" Silberman recalled. "You could have knocked me down with a feather. But that's Clarence: private, knows what he wants, and goes ahead and gets it."

Lamp's sincere, down-to-earth manner quickly won over Thomas's family and friends, even those who, given his college radicalism, were surprised to learn that she was white. When Thomas was ultimately nominated to the Supreme Court four years later, however, some black critics publicly questioned the match, even suggesting that it symbolized a rejection of his "blackness."

With the blessing of her parents, Lamp married Thomas at her home church in Omaha on May 30, 1987. Guests arriving at the ceremony passed a small table holding photographs of the soon-to-be married couple, a discreet announcement that the groom happened to be black.

Thomas returned to Washington after the wedding and prepared to give a major speech at the Heritage Foundation, perhaps the most respected and influential of Washington's conservative think tanks. For some, speaking to a Heritage audience was the political opportunity of a lifetime: the chance to prove conservative credentials, impress political allies, and curry favor with the GOP establishment.

That was not the speech Thomas intended to give. Instead, he readied a blistering indictment of the Republican Party's shabby treatment of black conservatives. The Clarence Thomas who appeared at Heritage on June 18 was a far less idealistic man than the one who had impressed Juan Williams seven years earlier, brimming with optimism. After seven years of bickering over quotas, affirmative action, and busing, Thomas was banged up and disenchanted. He had a few things to get off his chest.

In the speech he criticized the sharply negative tone of the Reagan administration toward civil rights, calling the administration's decision on Bob Jones University a "fiasco." He chided the administration for taking so long to support renewal of the Voting Rights Act. Black conservatives such as himself were under constant pressure to prove their conservative credentials, Thomas said. At best, they were treated with indifference by white conservatives, he said, and more often with suspicion.

"It was made clear more than once that, since blacks did not vote 'right,' they were owed nothing," Thomas said. "For blacks the litmus test was fairly clear. You must be against affirmative action and against welfare. And your opposition had to be adamant and constant, or you would be suspected of being a closet liberal."[1] Thomas, of course, was opposed to both affirmative action and welfare, but he resented the idea that black Republicans had to prove their racial politics before gaining acceptance for their conservative ones.

The Republicans' failure to attract blacks, Thomas said, was a fail- ure of principle. The party treated blacks like just another interest group, unworthy of pursuit. "Polls rather than principles appeared to control," he said.

Thomas hoped the speech would be discussed and remembered. And it was. But for altogether different reasons.

Nine days after Thomas's speech, Supreme Court justice Lewis F. Powell, Jr., a moderate, announced his resignation. On July 1 Reagan nominated Robert Bork to fill the vacancy, describing him as an open- minded jurist who would provide scholarly balance to the Supreme Court.

Bork's nomination mobilized many of the same interest groups that would become involved in Thomas's Supreme Court nomination four years later. The ferocity of liberal opposition to Bork rallied con- servatives, hardening their resolve to remake the Court in a conserva- tive image. Bork's ultimate defeat emboldened liberal advocacy groups to take on other conservative judicial nominees. The intensity of the Bork confirmation battle on both sides of the aisle left bitter feelings in Washington for years to come.

The tone of the anti-Bork campaign was set within minutes of Reagan's announcement. Rushing to the Senate floor, Senator Edward M. Kennedy of Massachusetts declared: "Robert Bork's America is a land in which women would be forced into back-alley abortions, blacks would sit at segregated lunch counters, rogue policemen could break down citizens' doors in midnight raids, schoolchildren could not be taught about evolution, writers and artists could be censured at the whim of government."[2]

Kennedy's hyperbole was calculated. He wanted a shrill denunciation of Bork's record to scare off other senators from endorsing Bork too quickly, buying the opposition time to mount a credible campaign. All through the summer a diverse coalition of liberal organizations, organized labor unions, and women's groups mounted a grassroots campaign against Bork. The campaign generated fierce opposition to the nominee among women who feared Bork would overturn *Roe* v. *Wade* and civil rights supporters who believed he was hostile toward minorities. Thousands of letters and calls came in to Capitol Hill. Protesters flocked to Washington during the confirmation hearings in September. Law-school deans and professors from around the country urged the Senate to reject Bork. Even the American Civil Liberties Union, normally neutral in Supreme Court battles, produced a report critical of Bork's record on civil liberties.

The hearings opened September 15. The same day, William Coleman, a lifelong black Republican, opposed Bork in a *New York Times* editorial. The piece signaled the extent to which Bork was opposed among black civil rights leaders, including some Republicans. Thomas was virtually alone among black Republicans in supporting Bork, who had taught at Yale while he was a law student. "Judge Bork is no extremist of any kind," Thomas said in August. "I am appalled at the mud-slinging cum debate over the Bork nomination."[3]

During his confirmation hearings, Democratic senators effectively put Bork on the defensive, peppering him with questions about civil rights and abortion. Bork's writings and speeches over a twenty-year legal career provided ample fodder. The strategy, forcefully articulated by Laurence Tribe two years earlier, was to make Bork's judicial philosophy the issue.

Bork, however, also proved to be his own worst enemy. He rolled his eyes at questions from the Judiciary Committee. He sighed loudly.

No amount of diplomacy concealed his deep disdain for his Senate inquisitors, and the arrogance cost him support from potential allies.

Bork testified for thirty-two hours over five days. More than 110 witnesses testified. The Senate Judiciary Committee voted 9 to 5 against his confirmation, sending the nomination to the full Senate without a recommendation. In the Senate, where Democrats held a majority, Bork's chances looked bleak. On October 23 the Senate voted him down, 58 to 42, the largest margin of defeat for any Supreme Court nominee. Bork returned to the D.C. Circuit Court of Appeals to lick his wounds, an angry and bitter man. Less than three months later, in January 1988, he announced his resignation, beginning a chain reaction that would ultimately lead to Clarence Thomas.

By the close of the Reagan administration in January 1989, Thomas was reaching the end of his tenure at EEOC. One of his final tasks was to see the agency safely to a new headquarters. The State Department was claiming the EEOC's building on E Street, and the Reagan administration was proposing to move the headquarters to the Washington suburbs. Thomas vehemently opposed moving the EEOC out of Washington. He thought it important symbolically that it remain downtown, accessible to people who really needed it. With the help of Democrats on Capitol Hill, notably Congressman John Lewis of Georgia, Thomas won the fight. A new headquarters building was planned on Washington's L Street, just a few blocks from the White House.

Meanwhile, a major scandal was brewing at the EEOC. Through a misinterpretation of the law, the EEOC field offices had allowed thousands of age-discrimination cases to expire without action. "We blew it," an apologetic Thomas told Hugh Downs on ABC's *20/20* program in July 1988. Democrats on Capitol Hill requested three years of EEOC documents relating to the agency's handling of age discrimination cases and subpoenaed Thomas to appear before the

Senate Aging Committee. As the number of potentially mishandled cases mounted, Congress enacted special legislation to allow the plaintiffs whose cases had lapsed to renew them with the EEOC.

While he accepted responsibility for the fiasco, Thomas was bitter about the negative publicity, believing his enemies were using the controversy to taint his entire eight-year record. A "personal vendetta" was how he described it.[4] It was a lousy way to finish his EEOC tenure.

Thomas vented his frustration in a series of speeches that attacked decision making in Congress. Congress had ceased to become a deliberative body, he told the Palm Beach Chamber of Commerce in May, and was acting more like a second division of the executive branch. It exerted control over policy and decision making through committee, forcing changes that were often at odds with congressional intent. "Congress as a body no longer deliberates or legislates in a meaningful manner, because members are more concerned with other more direct, more profitable, less risky, and less open methods of rule," Thomas said.[5]

Thomas used Title VII of the Civil Rights Act, the provision governing nondiscrimination in the workplace, as an example. He noted that the act ordered the EEOC to protect individual victims of discrimination. Congressional committees, however, at the urging of the civil rights community, had transformed the act into a vehicle that rewarded entire classes of victims. The act's expressed concern for the protection of individual rights had been transformed into a mandate for group remedies, Thomas said.

Thomas also faced the challenge of finding new employment in 1988. Early in the campaign year, it appeared that Democrats were mounting a strong bid to regain the White House after eight years of Republican control. The optimism peaked at the Democratic National Convention that summer, when the party nominated Michael Dukakis, a Massachusetts governor, as its presidential candidate. Republicans gave the nod to Vice President Bush.

Thomas considered opportunities in the private sector. He talked to the American Bar Association about becoming its national director. Pam Talkin, Thomas's chief of staff, recalled that Thomas used to dream about managing a larger government agency, one where he could make a mark without generating so much controversy. He loved the idea of running the General Services Administration, the plain vanilla agency charged with providing management support to other government departments. "Pam, we could run GSA and make the government work better," he told Talkin. "We'd stop being so petty and so bureaucratic."

As the nation prepared to go to the polls in November, Senate Democrats shelved fifteen of Reagan's judicial nominees without a committee vote, including a forty-seven-year-old woman named to succeed Bork on the D.C. circuit. Democrats were banking on a new Democratic president to fill the vacancies instead.

On election day, however, Bush defeated Dukakis, extending Republican control of the executive branch for another four years. Bush's victory changed Thomas's employment picture dramatically. He was certain another appointment would now be forthcoming, recalled Linda Rule, his special assistant. Thomas's hunch, Rule recalled, was that he was headed for a senior post at the Department of Justice.

But when the White House called, it wasn't about Justice. President Bush wanted to nominate Thomas to take Bork's place on the D.C. circuit.

There was only one catch: Thomas didn't want the job.

CHAPTER 13

RELUCTANT JUSTICE

Clarence Thomas dozed fitfully before stumbling out of bed in the predawn darkness of October 11, 1991. In four hours he was scheduled to face the nation to testify about allegations that he had sexually harassed Anita Hill a decade earlier. The preceding week had been surreal. His confirmation to the Supreme Court had been assured before Hill's allegations were leaked to the media six days earlier. Now he faced another round of hearings and widespread predictions that his nomination was doomed. Reporters and photographers camped outside his suburban Virginia town house around the clock, making Thomas feel like a prisoner in his own home. He'd slept only an hour the night before, and no more than a few hours at a time all week. He'd finished writing his statement only an hour or two earlier.

Over the telephone at about six-thirty, Thomas read his statement to Senator Danforth, who suggested two changes. Afterward, Thomas joined his wife and friends in the living room. They held hands and prayed aloud, waiting for Thomas's ten o'clock appointment with the American people. Charles James, descended from three generations of Baptist ministers, told Thomas God had given him this trial to build his character and make him a better judge.

"Do you believe in God?" James asked him, Kay James recalled.

"Yes," Thomas replied.

"Then act like it."

U.S. marshals whisked Thomas and his wife to the Russell Office Building on Capitol Hill. Inside, hundreds of police, news reporters, and partisans, both for and against Thomas, jammed the white-marble corridors. The friendly faces included black churchwomen clad in bright green dresses, white conservatives in suits, and dozens of Thomas's former colleagues from the EEOC. The crowd clapped rhythmically and shouted, "Thomas! Thomas!"

Clarence and Virginia Thomas emerged from Danforth's office at around ten o'clock, squinting at the harsh television floodlights. Police formed a circle around the couple as they jostled awkwardly through the crowd. Thomas hurried up the marble staircase to the third floor and into the grandeur of the Russell Caucus Room, as news photographers clicked shutters in rapid fire.

Thomas, dressed in a blue suit, sat stiffly, folding his large hands over his typewritten remarks. He swirled his tongue every few moments to ease the dryness in his mouth, then clenched his jaw. Sweat beaded on his forehead as he stared impassively at the fourteen white men in front of him.

Television beamed Thomas's every move to millions of Americans around the country. The nation waited for him to respond to allegations that he had used vulgar language around his former staffer and had pressed her for dates. The alleged incidents occurred ten years earlier, yet official Washington talked of nothing else all week. Many were certain that Thomas's nomination was about to implode.

At that moment, only his wife and his mentor, Missouri senator John Danforth, knew what Thomas planned to say.

He spoke slowly, recounting how he had met Hill through his friend Gil Hardy, offered her a job at the Department of Education and another at the EEOC. He came to the sentence Danforth asked him to insert. "If there is anything that I have said that has been misconstrued by Anita Hill or anyone else to be sexual harassment, then I can say that I am so very sorry and I wish I had known."

Then he angrily described the agony of the preceding days. "During the past two weeks, I lost the belief that if I did my best, all would work out," Thomas said. "I called upon the strength that helped me get here from Pin Point, and it was all sapped out."

Suddenly, Thomas sounded like he was bowing out. "I'm not going to allow myself to be further humiliated," he said. "Enough is enough."[1]

That was it, thought Clint Bolick, one of Thomas's former EEOC staffers. Thomas was withdrawing.

Hill's allegations were less than a week old when Thomas began testifying, but the ingredients for the firestorm had been swirling in the nation's capital for years. Potent interest groups had transformed Supreme Court nominations into life-or-death battles. The combatants hoisted banners such as "our living Constitution" and "judicial restraint." But the terms of engagement were euphemisms for the underlying issues: abortion, school prayer, and affirmative action.

Relations between Democrats and Republicans on Capitol Hill were at new lows when President Bush nominated Thomas to the Supreme Court. By 1991, Newt Gingrich of Georgia and other House Republicans had already chased former Democratic Speaker Jim Wright from power over a questionable book deal. Senate Democrats blocked John Tower of Texas from becoming Bush's defense secretary over allegations of boozing and womanizing, angering many Republicans. Democrats were mounting a campaign to reclaim the White House after twelve years of GOP control, while Gingrich and House Republicans were attempting to seize power in Congress.

"Thomas became one act in a larger drama between two parties," said Ralph Reed, the former Christian Coalition director. "We were trying to get the House. They were trying to get the White House. And the politics of confirmation became a battleground."

Most significant, perhaps, Hill's allegations against Thomas combined race and sex, a combustible cocktail that transformed the confirmation hearings into an unprecedented public spectacle. Partisan acrimony over the handling of the allegations fueled the rancor that permeates judicial confirmations today. The aftermath ultimately engulfed President Clinton, feeding the determination of conservatives to bring him down in sexual scandal.

Yet Thomas's confirmation also burned out rotten wood from America's darkest thickets, exposing sexual harassment in the workplace and leading to fundamental changes in workplace relationships between men and women. Companies all across America wrote or rewrote guidelines on harassment. The hearings generated a groundswell of political support for women in government, reshaping the Congress. Anger over the hearings also helped catapult President Clinton to the White House and drew millions of American women into the Democratic Party.

The principal players in the drama, Anita Hill and Clarence Thomas, paid for the historic changes with their reputations. The confirmation left a bitter taste in Hill's mouth and redefined the course of her young life.

Thomas would never be the same. He emerged scarred and traumatized, suffering public humiliation that took him years to overcome. He rebuilt himself, but he was profoundly altered. The confirmation changed where and how he lived, whom he trusted, what he said in public, even what he read. The exposure robbed him of the thing he prized the most: his good name.

Clarence Thomas had never aspired to be a judge. In the Savannah of his childhood there were no black attorneys, much less black judges.

As a libertarian, Thomas was also not comfortable with the idea of sitting in judgment of others. When he learned in early 1989 of President Bush's desire to nominate him to a circuit court judgeship, Thomas told the new administration he wasn't interested.

At forty-one, he wanted to consider his opportunities carefully. Appointments to the judiciary implied a lifetime commitment. The bench guaranteed a comfortable life but would not make Thomas a rich man. He also felt too young to be a judge. "I'm not serious enough!" he exclaimed to Linda Rule. He'd been thinking of buying a black Corvette. "How could I be a judge in a black Corvette?"

Thomas also wasn't eager to undergo another hostile Senate confirmation. The stakes would be even higher for the circuit court, and Thomas knew he would draw controversy.

The man driving his nomination to the U.S. Circuit Court of Appeals for the District of Columbia Circuit was C. Boyden Gray, Bush's White House counsel. John Sununu, Bush's chief of staff, said Gray had been talking up Thomas to Bush for years.

At Gray's request, Thomas delivered copies of his speeches and legal writings to the White House for review just weeks after Bush's inauguration.

Sick of Washington, Thomas said he remained ambivalent about the process. "I never made up my mind, and just sort of went along and filled out the forms," he said. "I figured it would die at some point and I'd be off the hook."

Laurence Silberman, a judge on the D.C. circuit, was among those counseling Thomas to take the appointment. He impressed on Thomas how independent judges were. No other position was as free from political pressure. The description appealed to Thomas, particularly after he had labored ten years in the political trenches.

Administration officials told the media in May 1989 that Bush intended to nominate Thomas for Bork's seat. Washington insiders noted the significance of the appointment because of the D.C. circuit's unofficial reputation as the proving ground for the Supreme Court; the court has graduated more judges to the Supreme Court than any other.

The Alliance for Justice, a key player in the anti-Bork campaign, quickly determined that Thomas was a judicial conservative, urging the Senate Judiciary Committee in an October letter to examine Thomas's "persistent and public attacks on affirmative action."[2] Ralph Neas, director of the Leadership Conference on Civil Rights, was another early opponent to Thomas's nomination and recruited others to stop it. Fourteen House Democrats opposed Thomas in a letter delivered to Bush in July, three months before the White House even sent the nomination to the Senate. The liberal coalition included two major labor unions, the National Organization for Women, the NOW Legal Defense Fund, and several organizations with older constituencies who were angry over the EEOC's handling of age-discrimination cases.

Notably absent from the coalition, however, were the nation's major civil rights organizations, including the NAACP.

The Senate Judiciary Committee voted to recommend Thomas for confirmation in February 1990 by a vote of 12 to 1. The lopsided margin, however, concealed Democrats' opposition to Thomas for any future appointment to the Supreme Court. Already there was speculation in Washington about Justice Thurgood Marshall, whose health was failing. Prior to the committee's vote, Senator Joe Biden of Delaware pointedly told Danforth that his support was "just for the court of appeals," Danforth recalled. Thomas was getting a pass but couldn't expect similar treatment again. On March 6, 1990, the Senate confirmed him on a voice vote. Only Senators Howard Metzenbaum of Ohio and David Pryor of Arkansas, both angry at Thomas for the lapsed age-bias cases at EEOC, voted no.

Thomas was nearly catapulted to the Supreme Court sooner than anyone in Washington imagined. Less than five months after he took office, Supreme Court justice William J. Brennan, Jr., a liberal icon, announced his retirement. President Bush wanted to nominate Thomas, Gray said. But he and Attorney General Richard Thornburgh argued that Thomas was too young and inexperienced. Bush reluctantly heeded the advice but indicated that Thomas was his presumptive first choice for the next vacancy, Gray said. Anxious to avoid controversy, the White House also didn't want another Robert Bork, said Sununu. "Everybody was trying to bring the process back to normal."

Instead, Bush nominated David Souter, a political unknown whom Bush had only recently named to the First Circuit Court of Appeals.

Souter's Senate confirmation displayed how much the White House had learned from the Bork battle. Justice Department lawyers coached Souter for weeks on how to behave before the committee, advising him how to deflect questions about controversial topics such as abortion and privacy rights. Souter also generated little controversy because liberal advocacy groups knew that he was sympathetic to some of their issues, based on intelligence gathered in his home state, said Nan Aron of the Alliance for Justice. "We had heard Souter would be good," she said.

Souter testified for three days, and the Senate confirmed him 90 to 9, a quick and painless affair that would stand in stark contrast to Thomas's ordeal less than a year later.

Thomas loved his new job as one of twelve judges on the D.C. circuit. His days were no longer endless, back-to-back meetings and phone calls. Joining the court was like joining a monastery—a contemplative life where privacy and quiet were the norm. And, making nearly $133,000 a year, he could have his cake and eat it, too. Shortly after he started at the court he bought a black, limited-edition Corvette ZR-1, his second sports car after selling his first Camaro.

Thomas relied heavily on Judge Silberman for practical advice and philosophical counsel. Coming squarely out of the conservative mold, Silberman believed that the courts, following the lead of the Warren Court, were making law and ignoring the intent of the Constitution's framers. He believed the Constitution sometimes allowed flexibility in interpretation but urged Thomas to eschew flexibility in favor of a narrow interpretation.

Early on Silberman told Thomas that the art of judging was learning to think like a judge. When he decided a case, he could not do so as a father, spouse, or outraged citizen, he said. He told Thomas to ask himself what his role was as a judge. He must never ask himself: How do I want this case to come out?

Yale classmate Don Elliott lunched with Thomas in the fall of 1990 and found him thinking deeply about issues far removed from the law. "He was thinking about the problem of how, when a community had lost the norm of public service, could you go about restoring that?" Thomas also talked about his EEOC critics. "He bore them no rancor," said Elliot. "Here was a person who was truly able to forgive and understand his detractors."

On the Supreme Court, meanwhile, Thurgood Marshall's health continued to fail. The eighty-two-year-old walked with a cane. His eyes teared constantly, making it difficult for him to read. With Brennan gone, Marshall felt alone on the Court. He told friends he could wait no longer for Democrats to regain the White House. Marshall informed Bush of his retirement in a two-sentence letter on Thursday, June 27, 1991.[3]

The letter landed on Boyden Gray's desk at the White House around noon. Gray picked up the phone and called Thomas, who returned the call later that afternoon. "Are you ready to take another walk?" Gray asked.

Thomas met one of Gray's assistants on a street corner. The aide drove him to the Justice Department, where he was ushered into a soundproof situation room. Justice lawyers probed him for details that could present problems at his confirmation, asking him about his divorce and marriage to a white woman. There wasn't much left to ask. After four Senate confirmation hearings and five FBI background checks, they reasoned, any skeletons in Thomas's closet would have come out.

At about the same time, Sununu called Tom Jipping at Coalitions for America, one of the lead conservative organizations on judicial nominations, and asked him to fax over a list of his top picks for Marshall's seat. Thomas topped Jipping's list. "The entire conservative movement not only supports him, but believes in him," Jipping wrote. "No dissent is likely from anywhere in the movement."[4]

One by one, the stars were beginning to align in Thomas's favor.

The next morning Thomas was summoned to the White House to be announced as Marshall's successor. He was escorted into a holding room in the basement to await the signal to join the president for the announcement. Sitting alone, Thomas began writing out a few words of acceptance.

Somewhere above him, however, a disagreement had erupted over the timing of his nomination. Justice lawyers were urging the president to wait. Bush put off a final decision on Thomas and left for his summer retreat in Maine.

Thomas sat in the basement for several hours, unaware that his nomination had unexpectedly gone wrong. An aide to Gray finally broke the news. Thomas left the White House suddenly uncertain about his chances. Later in the day he learned that the Justice Department planned to interview Fifth Circuit Court of Appeals judge Emilio Garza, a former Marine who hailed from Bush's adopted state of Texas. Garza was just as young as Thomas but hadn't yet written any opinions as an appellate judge. A nomination to the Supreme Court would make him the first Hispanic justice.

Aboard *Air Force One*, Bush told reporters that he hadn't made up his mind about Marshall's successor. "I want to keep in mind representation of all Americans," he said.[5]

Ironically, Thomas's race was one of the biggest strikes against him in Bush's mind, according to Sununu and Gray. The president hated the idea that Thomas's nomination would appear to be a "quota," Gray said. "It looks like I'm putting a black in for a black," the president told his counselor. Yet Bush only seriously considered two men, both minorities.

Thomas said he felt relieved to have avoided another confirmation battle. The next day, a Saturday, while Garza was being interviewed at the Justice Department, Thomas and Virginia hopped in the car and drove to Annapolis on the Chesapeake Bay for a seafood dinner to celebrate having just dodged a bullet. "Better him than me," Thomas said of Garza.[6]

On Sunday Thomas was working at his office when a clerk burst in telling him the president was on the line from Maine. Thomas picked up the phone. Could Thomas come to Kennebunkport for lunch Monday to talk about that "Supreme Court thing"? Bush wanted to know.

"Yes, Mr. President," Thomas stammered.

The conversation was brief. Bush asked Thomas to give the job some thought, but he stopped short of telling him that he planned to nominate him. Hanging up the phone, Thomas believed the invitation to Maine was for an interview.

The following morning he flew to New Hampshire, a destination chosen to avoid media detection. Sununu met him at the airport and drove with him to Kennebunkport. Thomas was tense and nervous, recalled Sununu, who kept assuring him that everything would be okay. As the car pulled up to the Bush compound, Sununu warned him that the media was going to identify them. Sununu handed his sunglasses to Thomas. "So they'll think you're me," he said with a straight face. Thomas stared at Sununu for a full half minute before filling the car with guffaws.

Thomas arrived at the Bush home while lunch preparations were under way. He saw the First Lady before he saw the president. "Congratulations!" Barbara Bush exclaimed, then apologized for letting the cat out of the bag. Thomas's heart sank, he recalled later. "From then on it was surreal," he said. "It was an out-of-body experience."

Bush escorted Thomas to his bedroom, the only quiet place in the house, and asked him two questions. Could he and his family withstand the pressure of "the bruising fight ahead"? Thomas said yes. Using a baseball analogy, the president then asked Thomas if he could "call 'em like he saw them." Thomas said yes again. "That's the way I've lived my whole life," he said. Bush shook his hand and told him he would introduce him as Marshall's successor after lunch.[7]

The sun was shining gloriously as Thomas stepped outside with President Bush and faced the media. It was the first of July. A warm breeze blew in from the Atlantic Ocean, and the color of the sky was a deep, brilliant blue. Introducing Thomas, Bush noted that Thomas had been his first appointment to the D.C. circuit. "I believe he'll be a great justice," Bush said. "He is the best person for this position." Bush ticked off Thomas's educational background and work experience. "He has excelled in everything that he has attempted," said Bush. "He is a delightful and warm, intelligent person, who has great empathy and a wonderful sense of humor. He's also a fiercely independent thinker with an excellent legal mind."

Bush stepped aside, and Thomas stood behind the wooden podium, decked with the blue-and-white presidential seal. "As a child I could not dare dream that I would ever see the Supreme Court, not to mention being nominated to it," Thomas said. "I thank all of those who have helped me along the way, and who have helped me to this point and this moment in my life, especially my grandparents who are—" He stopped midsentence, momentarily unable to continue. His eyes welled up with tears. He cleared his throat before continuing. "Especially my grandparents, my mother, and the nuns, all of whom were adamant that I grow up to make something of myself."

The thrill of the moment lasted less than twenty-four hours.

. . . .

The National Abortion Rights Action League (NARAL) drew first blood. Within hours of the nomination, its lawyers were scouring Thomas's writings and speeches, looking for a position on abortion. They believed Thomas was anti-choice, but they needed proof.

They thought they found it in a four-year-old speech at the Heritage Foundation, the one in which Thomas faulted the Reagan administration for its blunders in civil rights policy.[8] At the tail end of the speech Thomas praised Lewis Lehrman, a noted New York conservative, for an essay that used natural law to provide a moral basis for laws outlawing abortion. Lehrman's essay on the "Declaration of Independence and the meaning of the right to life is a splendid example of applying natural law," Thomas had said. Thomas made no direct reference to Lehrman's antiabortion arguments, or to *Roe* v. *Wade,* which the Lehrman article also criticized. Thomas inserted the reference as a gracious nod to the sponsors of his speech. Lehrman was a trustee of Heritage, and the auditorium in which he spoke was named for him.

Seizing on the Lehrman reference, executive director Kate Michelman said NARAL would oppose Thomas unless he repudiated his anti-choice statements and recognized a right to abortion in the Constitution. "This is concrete evidence that Judge Thomas has not only passed the Bush administration's litmus test [on abortion], he's roaring out there," said Michelman on July 2. "It's clear that where he's headed is extraordinarily dangerous."[9]

At the time, abortion-rights advocates had reason to worry about the Supreme Court. Marshall and Brennan were two of the Court's strongest pro-choice justices. Abortion rights groups believed there were already four votes to overturn *Roe.* Thomas would be the decisive fifth vote, they feared. Their fears had grown just five weeks earlier, when Justice Souter joined a 5 to 4 majority that withheld federal funds to family planning clinics that counseled women on abortions, the Republican appointee's first vote on abortion.

A direct challenge to *Roe* was more than a theoretical possibility. An abortion case from Pennsylvania was likely to land on the Court's docket in the term that would begin in October. The fate of *Roe* hung in the balance, abortion-rights advocates believed.

Passions were running into the red zone when the National Organization for Women (NOW) met in New York days after Thomas's nomination. NOW president Patricia Ireland vowed to throw the organization's resources into defeating Thomas, declaring that NOW had already begun phoning members.

Yet the remark that made news came from Flo Kennedy, an outspoken feminist known for her incendiary comments. "I'm embarrassed as a black person that they ever found this little creep," the seventy-five-year-old Kennedy said of Thomas. "Where did he come from?" Then she added: "We're going to Bork him," she said. "We need to kill him politically."[10]

Kennedy's comments amounted to a declaration of war and crackled on conservative communication lines around the country. People who had never even heard of Clarence Thomas were suddenly volunteering to help him. Money flowed into conservative Washington organizations to pay for the campaign to support Thomas.

"Everyone wanted desperately to have a conservative justice," said Clint Bolick, who was then launching a libertarian think tank that would join the pro-Thomas campaign. "There was also a huge reservoir of animosity over Bork. A lot of people were not there because they loved Clarence Thomas but because of Robert Bork. For them, it was payback time."

Ricky Silberman worked with both Tom Jipping and Bolick to rally supporters, developing a strategy to match the opposition tit for tat. During Bork's confirmation, liberal organizations had mounted street protests and demonstrations on Capitol Hill. Conservatives planned their own "street theater," said Jipping, arranging to bring in conservative ministers and church people to show their support of Thomas. The group also resolved that no charge against Thomas would go unanswered.

Within weeks, two conservative organizations were producing a sharply negative television ad that attacked Senator Edward M. Kennedy and two other Democrats who were prominent in the anti-Bork campaign, signaling an aggressive defense of Bush's nominee.

At the White House, the administration assembled a team to prepare Thomas for his confirmation hearings and lock up Senate votes. Ken Duberstein, a former Reagan chief of staff and now a Washington lobbyist, was put in charge of public relations. He was the architect of what became known as the "Pin Point strategy," a campaign to package Thomas to the public as an American success story in the Horatio Alger mold. A young Justice Department lawyer, Michael Luttig, was in charge of preparing Thomas for legal questions. Fred McClure, the White House congressional liaison, was the vote counter. In the Senate the White House relied on Danforth. But the administration also had a powerful ally in Orrin Hatch, the reedy Mormon from Utah. Hatch dressed like an early twentieth century dandy, with striped shirts and gold tie clasps. One of the shrewdest political minds in the Senate, Hatch was the conservative counterpoint to Kennedy.

Danforth and McClure immediately pushed for confirmation hearings before the Senate's August recess to limit the time for opponents to organize, an effort that ultimately failed. Duberstein put Thomas and Virginia under strict orders not to talk to the media on any subject. As with Souter, the Justice Department prepared Thomas to step gingerly around topics such as abortion, believing that Bork had been too blunt.

Among the group, however, Thomas was the most astute student of the Bork nomination. He studied an analysis of the Bork battle entitled *The People Rising: The Campaign Against the Bork Nomination.* The book's central thesis was that Bork united a powerful lobby of civil rights organizations, women's groups, and liberals that generated a groundswell of public opposition to his nomination. For Thomas the strategy was clear: He had to break the liberal coalition in half. Since he was black, Thomas's best shot lay with splintering the civil rights organizations.

Throughout early July he waged an intense, behind-the-scenes lobbying campaign on his own behalf. The White House enlisted the help of black members of the administration, including Dr. Louis Sullivan, Bush's secretary of the Department of Health and Human Services. Sullivan, a native of Georgia, was close to key civil rights figures, including Dr. Joseph Lowery, head of the Southern Christian Leadership Conference.

Sullivan arranged for Thomas to meet Lowery at a Washington town house. As Lowery recalled: "He looked me right in the eye and said, 'You'll be proud of me. I believe in the things that you and Dr. King believe in.'" Thomas also discussed his views on affirmative action, Lowery said, acknowledging that he had benefited from racial preferences in his own life. "He said he believed in affirmative action." (During the same time period, Thomas also met with Sharon McPhail, the president of the National Bar Association. Thomas told her that he believed in affirmative action remedies such as hiring goals and timetables, but only in cases where employers had been convicted of systemic discrimination, McPhail said.[11])

Based on the meeting with Thomas, Lowery and his committee from the SCLC decided not to oppose the nomination, a crucial early victory for Thomas. "Many of us figured this would be our last chance to get a black on the Court," said Lowery. "We thought if he gets in, then maybe some of us could talk to him."

Thomas also secured a commitment from the National Urban League, another prominent black organization, to remain neutral. He faced a far more difficult challenge, however, with the granddaddy of all civil rights organizations, the NAACP, which was already engaged in a power struggle over the nomination.

At the NAACP's annual convention in early July, Dr. Benjamin Hooks delayed a decision on Thomas until they could conduct a more extensive review of his record. Hooks said he also wanted to hear from the nominee himself, and Thomas was invited to meet with a small group of NAACP board members in July.

Better than anyone perhaps, Hooks understood how Thomas's nomination had divided the black community. Black leaders were torn between their desire to elevate another black American to the Supreme Court and their uneasiness with Thomas's conservative views on affirmative action. Since the historic *Brown* ruling, civil rights leaders believed no arm of the federal government had done more to safeguard the rights of black Americans than the Supreme Court. No matter what Thomas's politics, a black perspective on the Court was vital, many argued.

Many civil rights leaders were also convinced that if Thomas was defeated, President Bush would not nominate another black. The administration had given the civil rights community ample reason to fear such an outcome: The names leaked to the media before Thomas was announced included at least three Hispanics and one white woman. If the president was thinking about another ethnic minority, it was a Hispanic nominee, not a black one.

Throughout July, Thomas also concentrated on the second prong of the strategy to win confirmation, locking down the votes in the Senate. With Danforth as his constant companion, Thomas paid courtesy calls to senators on Capitol Hill. The meetings played to one of Thomas's strongest suits, his personal charisma and warmth.

From the beginning, the Senate's demographics tilted in Thomas's favor. Although Democrats controlled the chamber, fifty-seven to forty-three, the number included a dozen southern Democrats, who tended to be more conservative and also represented large black constituencies.

By the August recess, Thomas had personally met with sixty senators and was fast approaching the fifty votes he needed to secure confirmation. McClure was increasingly optimistic that Thomas would sail through the Senate, provided he performed well in the confirmation hearings.

While Thomas lobbied senators, however, the anti-Thomas coalition was also hard at work. Key activists urged Democratic senators to withhold judgment until after the confirmation hearings. The strategy was to freeze Democrats from endorsing Thomas until the coalition had time to build its case against him.

With the exception of the NAACP and the other black organizations, the alliance that quickly assembled was the carbon copy of the anti-Bork coalition. It included the major women's and feminist organizations, several large labor unions, and gay-rights groups—all potent constituencies of the Democratic Party. Like the White House, the anti-Thomas coalition believed the key to victory lay with conservative southern Democrats and moderate Republicans.

The organizations threw themselves into building public opinion against Thomas. Even before he was nominated, liberal organizations had been sifting through his speeches and writings for ammunition. They found plenty. There were Thomas's statements in support of Robert Bork and others lamenting Oliver North's treatment in Congress during the Iran-contra investigation. Thomas's criticisms of congressional oversight of the executive branch became evidence of "Congress bashing" and his "disdain for the rule of law." His musings on natural law were a gold mine, liberals believed, because natural law was not a recognized method of constitutional interpretation, even among conservative judges.

Everything Thomas had said became grist for news stories. A 1983 speech in which Thomas said he "admired" Louis Farrakhan for his belief in self-help for black Americans allowed opponents to question whether Thomas was anti-Semitic. For Thomas, the story was particularly galling because most of his senior staff members at the EEOC had been Jewish women, and he was close to his Jewish colleague Judge Ruth Bader Ginsburg at the court of appeals. The Farrakhan story made *The New York Times,* along with other major papers, and required Thomas to issue a strong statement of denial.[12]

Thomas was prepared for the scrutiny of his published speeches and writings, but liberal organizations didn't stop there. They examined every facet of his life, combing through his divorce papers and his meager property and financial holdings, even researching the Episcopalian church he attended with his wife in Virginia. Some church members had opposed a local abortion clinic in the mid-1980s. Newspapers wrote stories about the church, even though there was no suggestion that Thomas had participated in any of its antiabortion activities. The anti-Thomas coalition requested memos and documents from the EEOC under the Freedom of Information Act.

By 1991, the tactics were part and parcel of any campaign against a member of Congress or candidate for the White House. Never before, however, had they been employed against a nominee for the Supreme Court.

"They were searching like mad," said journalist Juan Williams. "People were calling me at my desk and asking, 'What have you got on Clarence Thomas?' They wanted him."

Researchers conducted scores of interviews with EEOC employees. They asked about Thomas's management style, his treatment of employees, his role in specific cases. Some former employees said Thomas didn't listen to advice and seemed to make up his mind ahead of time—observations that were factored in to conclusions about his judicial temperament. None of the former EEOC employees, however, turned over the name Anita Hill.

Hill's allegations first surfaced at a Washington dinner party in July, according to the Alliance for Justice's Nan Aron, who learned of the information shortly thereafter. Wanting no direct involvement with Hill, Aron passed her name and a rough outline of her allegations to the staff of Senator Metzenbaum, who had opposed Thomas for the D.C. circuit. Metzenbaum's staff, however, did nothing with the tip.

Meanwhile, the anti-Thomas coalition worked on a "message memo" outlining the strongest arguments against him. The work was tricky. Abortion-rights advocates wanted abortion to be the major club against Thomas. Civil rights leaders were more concerned about his positions on affirmative action. Labor leaders wanted a strong anti-worker theme.

In late July the coalition conducted focus groups on Thomas in Atlanta and Philadelphia to help craft the message. The locations were strategic. Philadelphia was Senator Arlen Specter's hometown and was close to Delaware, Judiciary Committee chairman Joe Biden's home state. Atlanta was in Thomas's home state.

In each location the pollsters held separate focus groups—one white and one black. The participants were shown a brief video on Thomas's career and conservative political beliefs, with special attention to his opposition to affirmative action. Pointedly, the montage included a picture of Thomas with his wife, a white woman. It also highlighted what Thomas said in 1980 about his sister, Emma Mae, and her dependence on welfare.

Black focus-group participants were sharply divided over Thomas. Some saw him as a political opportunist who had used affirmative action to get ahead, then turned his back on those left behind. But they also wanted a black on the Supreme Court.

The more white focus-group participants heard about Thomas and his positions on affirmative action, however, the more they liked him.[13]

The results presented a major problem for Thomas's opponents: His opposition to affirmative action, likely to build antagonism toward him among black Americans, was the very issue that made him attractive to whites.

But the white focus-group members handed pollsters a different issue: They said they assumed Thomas's only qualification was his race. The anti-Thomas coalition seized on that assumption to craft the primary attack against the black nominee: Thomas wasn't qualified to serve on the Court. The subtext perpetuated a stereotype that blacks had been fighting for three hundred years.

The racially charged double standard had been in play ever since Bush nominated Thomas. The day of the announcement, journalists and pundits seized on Bush's assertion that Thomas was the "best person" for Marshall's vacancy. Presidential hyperbole was typical for such occasions, but Bush's remark in relation to a black man created an instant furor. Pundits mocked Bush's choice of words and all but declared Thomas incapable of serving on the court. The commentary also ignored Supreme Court history. Some of the Court's most distinguished jurists had had absolutely no judicial experience before joining the court; these included Louis Brandeis, Felix Frankfurter, William Douglas, and Earl Warren.

On July 31 the NAACP's board of directors met at a Washington hotel to vote on Thomas's nomination. Every board member had a seventy-two-page report on Thomas prepared by Wade Henderson, the Washington director. "Precisely because he is an African American, Thomas may be even more effective than a white conservative on the Court in legitimizing the attack and undermining the civil rights principles critical to African Americans," the executive summary concluded.[14]

Thomas's compelling life story, which captivated so many white Americans, carried no weight with NAACP board members. "Half of the people sitting there came from that type of background," said Percy Sutton, the former Manhattan Borough president and one of the board members. "All of us had been in the back of buses, or most of us. All of us had been arrested or most of us. There was nothing remarkable about him, no overcoming of hardship that impressed us."

Most argued that the board's vote should be unanimous, but Ben Andrews, the vice chairman, refused to bend to the pressure. "We were treating him differently than we treated other nominees with conservative views," said Andrews, the board's only Republican. The final tally was 50 to 1 against endorsement, with Andrews the lone dissent.

Two hours later the AFL-CIO issued a statement from Chicago opposing Thomas. With the NAACP on record against him, the giant labor organization felt safe to publicly oppose President Bush's black nominee.

Thomas saw the NAACP's action as an assault on his working-class roots and beliefs about self-reliance. Just as in his youth, Thomas viewed the NAACP leadership as an elite club of mostly fair-skinned, upper-class blacks who punished those who disagreed with them. The class blinders all but prevented Thomas from seeing any legitimate reasons the NAACP had for opposing him.

Following the NAACP's vote, the battle shifted to Indianapolis, where the National Bar Association was meeting. As the largest organization of black lawyers, the group, if it chose to back Thomas, could provide a counterpoint to the NAACP. The White House dispatched members of the administration who were NBA members to lobby delegates. Wade Henderson and Elaine Jones, with the NAACP Legal Defense Fund, worked the halls in opposition. Debate over the nomination raged for more than seven hours, the longest in the NBA's history. A federal judge from Washington State, Jack Tanner, championed Thomas on principle. A Carter appointee, Tanner bristled at the notion that the NBA should impose political litmus tests on judicial nominees. "We battled them to a standstill," he said. Thomas lost by four votes, 128 to 124.

The Leadership Conference on Civil Rights, meanwhile, launched the grassroots campaign against Thomas. Conference president Ralph Neas hired field coordinators in Pennsylvania and most southern states. Field staff were given samples of letters to the editor to send to local newspapers. "Thomas' paltry legal and judicial experience should make senators wonder what President Bush meant when he called Thomas 'the best man for the job,'" one sample letter read. The

coordinators were also responsible for developing a "shadow campaign" in each state, where activists were to appear at the political
events of targeted senators with anti-Thomas banners and literature.

On the other side, the Christian Coalition launched a round of ads
supporting Thomas in southern states that voted for Bush in 1988,
said Ralph Reed, its former director.

Thomas, meanwhile, was hunkered down in Washington studying
for the confirmation hearings. The Justice Department compiled fat
binders of Supreme Court cases and peppered Thomas with mock
questions. Thomas took to the books well, but Danforth worried
there was simply too much material to absorb. Thomas rose early in
the morning, sometimes by four o'clock, and began reading the daily
binders. He watched videotapes of David Souter's confirmation hearings, studying Souter's deft answers to Judiciary Committee questions.

Justice Department lawyers were particularly worried that Thomas
might come off too headstrong, according to Danforth. They baited
him with strong characterizations of his record and previous remarks
in speeches, hoping to discipline him into more measured responses.
They were also troubled by Thomas's responses to questions about
abortion and *Roe* v. *Wade.* He insisted he had no position on *Roe,* even
telling his handlers that he had never discussed the case with his wife.
"Oh, come on!" Duberstein said to Thomas during one session, according to Danforth. Despite their misgivings, Justice lawyers felt they
had no choice but to let the response stand.[15]

As the Senate's August recess drew to a close, Thomas received potentially damaging news from the American Bar Association. After an
exhaustive review of his record and writings, the ABA judged Thomas
"qualified" as opposed to "highly qualified" for the Supreme Court.
Justice Sandra Day O'Connor had been the last nominee to receive
the same ranking.

The ABA's evaluation team gave Thomas its highest rankings in integrity and judicial temperament but the middle grade for professional competence. The ABA committee noted that Thomas's legal

experience was primarily in the executive branch and said his published legal articles lacked analysis. The committee readers were more impressed with Thomas's appellate-court opinions. "His opinions are carefully and systematically reasoned, clearly articulated, respectful of the record (so far as we can tell), fair in consideration of opposing arguments, extensively supported by citations to authority, and demonstrate no obvious bias in decision."[16]

Thomas's opponents seized on the ABA's "qualified" rating to argue that Thomas wasn't fit for the job. "The country and the Court deserve better than a minimally qualified justice," Aron told *The Wall Street Journal*.[17]

The attacks on Thomas's competence, however, were beginning to splinter the anti-Thomas coalition. Henderson thought they fostered the stereotype that a black man wasn't smart enough to sit on the Supreme Court. He also reasoned that blacks would view attacks on Thomas's credentials as racism. Rather than diminishing Thomas's support in the black community, the attacks might only increase it.

Tensions were also building between civil rights leaders and NARAL, weakening the overall intensity of the campaign. Abortion was not the civil rights community's top priority, and they resented NARAL's influence over the direction of the campaign. The NAACP refused to sign off on the coalition's message memo because of its emphasis on both abortion and Thomas's qualifications, delaying the memo's completion for weeks. Civil rights leaders were at first opposed to having abortion-rights advocates part of the steering committee, and they balked at attending meetings at NARAL's Washington offices.[18]

Key Democrats from Georgia, South Carolina, and Louisiana, meanwhile, were refusing to commit against Thomas. Senator Sam Nunn of Georgia, a conservative Democrat, told Thomas's opponents that he would likely vote for his fellow Georgian, barring some extraordinary development in the hearings.

Polling timed to coincide with the opening of the hearings September 10 revealed that nearly two thirds of both black and white respondents had no opinion on Thomas whatsoever. Thomas was favored by 23 percent of black poll respondents and opposed by 15 percent; among whites, the figures were 26 percent for him and 10 percent opposed.[19]

As the hearings approached, Danforth and the White House believed Thomas was ready. Thomas bumped into Clint Bolick before the hearings began and gave him the thumbs-up sign. "Clint, I'm just going to tell it like it is and let the chips fall where they may," Thomas told Bolick. Increasingly, it appeared that only Thomas himself might sink his nomination.

CHAPTER 14

TRIAL OF HIS LIFE

Senator Howell Heflin of Alabama, a senior Democrat on the Senate Judiciary Committee, leaned over the dais and peered at Thomas through a pair of oversize spectacles. "You are somewhat of an enigma," Heflin drawled. "Some believe you are a closet liberal, and some, on the other hand, believe you are part of the right-wing extreme group."[1]

It was Friday, September 13, Thomas's fourth day of testimony. He had answered questions about abortion and deflected sharp questions about his conservative views. He had talked about growing up under segregation in Georgia and the instances where he felt the sting of bigotry. He had answered questions about his liberal years at Holy Cross, even his choice of combat boots and Army fatigues.

Heflin's pointed question captured the puzzle Thomas presented to the all-white committee. How could he be both black and conservative? For white liberals such as Heflin, black America was peopled by men such as Jesse Jackson of the Rainbow Coalition, Congressman John Lewis of Georgia, and Benjamin Hooks of the NAACP. Thomas simply didn't compute.

Democrats revealed the thrust of their attack from the very first round of questions. Biden grilled Thomas about natural law and quickly segued into abortion. He asked Thomas about rights to privacy in the Constitution and *Roe* v. *Wade*. Every Democrat on the committee repeated the same line of questioning, and every time Thomas delivered the same response. "The Supreme Court has made clear that the issue of marital privacy is protected, that the State cannot infringe on that without a compelling interest, and the Supreme Court, of course, in the case of *Roe* v. *Wade* has found . . . a fundamental interest in a woman's right to terminate a pregnancy," Thomas told Biden on the first day. But he stopped short of saying whether he believed abortion should be a constitutionally protected right, saying, "I do not think at this time I could maintain my impartiality as a member of the judiciary and comment on that specific issue."

Thomas's answers were almost a carbon copy of Souter's responses a year earlier. But the intensity of the abortion issue was now far greater. Women's organizations believed Democrats had allowed Souter to wriggle out of questions too easily. Democratic senators were far more dogged with Thomas, returning to abortion more than seventy times over the course of the hearings.

Exasperated with Thomas's responses, Senator Patrick Leahy of Vermont asked Thomas if he hadn't discussed the 1973 *Roe* decision with his Yale classmates.

"We may have touched on *Roe* v. *Wade* at some point and debated that," Thomas said. "I cannot remember personally engaging in those discussions. The groups that I met with at that time during my years in law school were small study groups."

"Have you ever had discussion of *Roe* versus *Wade* other than in this room?" Leahy continued, prompting laughter in the hearing room. "In the seventeen or eighteen years it's been there?"

"If you're asking me whether or not I've ever debated the contents of it, the answer to that is no, Senator."

Thomas also faced tough questions from Senators Kennedy and Metzenbaum, his old nemeses from his EEOC days. Kennedy quoted a 1987 speech in which Thomas described his views on socialism and the trend to equate individual freedoms with entitlements. "Which entitlements were you referring to as socialism?" Kennedy demanded. "Social Security or Medicare or Unemployment Insurance?"[2]

Democrats largely avoided Thomas's clearly defined views on affirmative action. Indeed, Thomas took the most pointed questions on affirmative action from a Republican moderate, Senator Arlen Specter of Pennsylvania. Thomas answered Specter's questions but seemed reticent to launch into forceful explanations. "The line that I drew was a line that said that we shouldn't have preferences or goals or timetables or quotas," Thomas said. "I felt it important that, whatever we do, we do not undermine the dignity, self-esteem, and self-respect of any group that we are helping."[3]

Many of Thomas's answers were as frustrating to his supporters as they were to his critics. He seemed too cautious, too programmed. Where was the feisty Clarence Thomas, the independent thinker who reveled in the give-and-take of intellectual debate? "He had been totally coached out of saying anything he thought," Bolick said.

Thomas was also clearly frustrating Democrats. "The vanishing views of Judge Thomas have become a major issue in these hearings," said an annoyed Kennedy on the fifth and final day of Thomas's testimony.

That afternoon, September 16, Thomas finished his testimony. He had testified for twenty-five hours, the second-longest grilling of a Supreme Court nominee in history. The remainder of the hearings would include panels of witnesses in support of or against his nomination. But the real drama was unfolding behind the scenes.

While Thomas was preparing to testify the week before, a top aide to Senator Metzenbaum was on the telephone with Anita Hill. James Brudney, a Yale classmate of Hill's, wanted to know details of Thomas's conduct toward her ten years earlier. Hill described unwelcome sexual advances from Thomas but told Brudney she did not want to testify and was unsure if she wanted to publicize her allegations to the committee.

Hill's contact with Brudney was just the beginning of a complex series of negotiations aimed at forcing her to put her allegations against Thomas on the record. The cast of characters extended far beyond Hill herself and included senior Democratic staff members not even on the Judiciary Committee, liberal lobbyists active in the anti-Thomas campaign, and sympathetic college professors. Hill's name and her allegations were Washington cocktail conversation weeks before she became a household name. In the end Hill made her own decision, but outside pressure had been building steadily as the hearings approached.

Following her conversation with Brudney, Hill told the staff of Judiciary Committee chairman Joe Biden that she might make her allegations if she could remain anonymous, explicitly telling them she did not want Thomas informed of them. Biden said no deal.

Hill's insistence on anonymity annoyed Biden. Initially, he refused to take a call from Hill because she wouldn't identify herself. "She went from wouldn't identify herself to identifying herself but unwilling to allow her name to be used, unwilling to testify, to willing to allow her name to be used, to writing a letter to senators on condition that I could guarantee that her comments would not become known," Biden told Specter, according to Specter.[4]

That same week, one of Strom Thurmond's staff members overheard Biden's chief of staff tell Biden that a certain female witness could "testify behind a screen."

"That's ridiculous," Biden said. "You can't do that. This isn't the Soviet Union."[5]

Over the next ten days Hill exchanged numerous phone calls with Democratic committee staff. She told Brudney she doubted she would be believed, but Brudney told her he was confident she would be treated credibly. Biden's staff also interviewed Susan Hoerchner, a

California judge, whom Hill said she had told of the harassment when the two were living in Washington in the early 1980s. Hoerchner backed Hill but also told committee staffers she was reluctant to go public.

The second phase of Thomas's hearings resumed September 17 with more than ninety scheduled witnesses. The chairman of the American Bar Association testified about Thomas's judicial qualifications. Kate Michelman, NARAL's director, testified about abortion and called Thomas a threat to reproductive freedom. Thomas's retired eighth-grade teacher, Sister Virgilius, spoke about the determined student she once knew. Guido Calabresi, dean of the Yale Law School, described Thomas as a diligent student who demonstrated strong capacity for personal growth. The attacks on Thomas's qualifications struck him as "liberal bigotry." "Just because he's black, people think 'stupid,'" Calabresi said later, describing his motivations for testifying. "I knew that he was smart."

Black witnesses disagreed sharply over Thomas. For some, he was a maverick thinker, a proud legacy of Booker T. Washington and the drive for self-determination. For others he was an Uncle Tom, who legitimized racism. Patricia King, a Georgetown professor, criticized Thomas for his statements about his sister, noting the opportunities he'd had were not open to her as a black woman. "All of us who have 'made it' have an obligation to help others, and to recognize that others need our help," King said: "Judge Thomas has been able to dream and to reach for his dreams. Yet, he ignored the need for or worked to deny that choice to others."

The Reverend Benjamin Hooks captured the nub of the opposition to Thomas among liberal black leaders with a forceful defense of group remedies to discrimination. "When I came along in 1949 and was admitted to the practice of law, there was not a single black in the courthouse except janitors and maids and one messenger," said Hooks.

"There were no blacks in the banks receiving money or using computers or typewriters, as the case might be. There were no blacks working in the stores downtown. Affirmative action has benefited America and millions of black people who otherwise would not have those jobs."[6]

Thomas's supporters defended his independence.

"Clarence Thomas does not fall conveniently into the liberal Democratic tradition that many members of the Black Caucus have defined for black Americans," said Robert Woodson, part of the small band of black conservatives in Washington. "There has been a gag rule imposed on the black community over the past twenty years that unless you see life through the prism of a liberal Democrat ... you will be suspicious, you will be castigated, and so that I think Clarence Thomas, because he does not espouse that position, is castigated."

William T. Coleman, Gerald Ford's secretary of transportation, declined to testify against Thomas, as he had testified four years earlier against Robert Bork. John Hope Franklin, the eminent scholar of American history, bowed out because of poor health. But Franklin also told the anti-Thomas coalition that he was upset that the hearings, as a whole, had dealt so little with civil rights.

On Friday afternoon, September 20, Biden gaveled the confirmation hearings to a close, promising a committee vote on the nomination within a week. The eight days of testimony were the third-longest in United States history. The most grueling days, however, were still to come.

The day before the hearings ended, Anita Hill told Democratic staffers that she wanted committee members to know about her allegations. Biden's staff told Hill the next day that her allegations would be turned over to the FBI and that the Bureau would interview her, Thomas, and other witnesses. Hill wanted time to think it over, telling Biden's staff she wished to consider the "utility" of involving the FBI.[7]

To help make up her mind, Hill consulted a Georgetown law professor, Susan Ross, recommended by Brudney. Ross suggested Hill consider writing a statement. So did Judith Lichtman, the head of the Women's Legal Defense Fund, whom Ross called with Hill's permission.

On Monday, three days after the hearings were over, Hill faxed a four-page statement to the Judiciary Committee. She described how her working relationship with Thomas was initially "relaxed and open." Three months into the job, she said, Thomas asked her out socially, and she said she declined.

Thomas persisted, Hill alleged. The relationship became strained, she said, when Thomas began discussing sex at work, using graphic language to describe his interests, and images he had seen in pornographic movies. The descriptions involved women having sex with animals, group sex, and rape scenes, Hill said. Hill told him the talk made her uncomfortable. She said she tried to change the subject. "However, I sensed that my discomfort with his discussions only urged him on, as though my reaction of feeling ill at ease and vulnerable was what he wanted."

Thomas stopped the behavior for about six months, Hill said. In the meantime, he was nominated to become EEOC chairman. "I felt a good deal of relief and hope that our working relationship could resume to what I considered to be a cordial, professional one," she said. Hill followed Thomas to the EEOC in the early summer of 1982. Three to five months later, she said, Thomas again asked her out and began making "random" comments about her appearance. Hill said she was feeling stress at work by late 1982 and began looking for another job in early January. She said she was hospitalized for five days in February with acute stomach pain, related to stress.

Before she left the EEOC in 1983, Hill said, she went out to dinner with Thomas. "He said that if I ever told anyone about his behavior toward me that it could ruin his career," Hill said. She said that since then she had had little contact with Thomas.[8]

Biden's staff gave Hill's statement to the staff of Senator Strom Thurmond, the ranking Republican on the Judiciary Committee, and notified the Justice Department and the White House counsel that an FBI investigation was now warranted. That night two FBI agents interviewed Hill at her home in Norman, Oklahoma.

At the White House Boyden Gray and his staff pored over Hill's statement. It seemed "cooked," according to Lee Liberman, one of Gray's assistants. Liberman was troubled that a ten-year-old allegation had suddenly materialized at the eleventh hour. Why hadn't Hill come forward before? Why had she followed Thomas to the EEOC if his behavior made her feel so uncomfortable?[9]

The next day FBI agents interviewed Susan Hoerchner, who corroborated Hill's account. Agents also interviewed two women who had worked with Hill and Thomas at the EEOC. They said they saw no inappropriate conduct from Thomas.

Wednesday morning Liberman called Thomas and told him to arrange for an interview with the FBI. He did not explain the reason.

"I felt like throwing up," Thomas told Danforth.[10]

At one o'clock that afternoon two FBI agents arrived at Thomas's home. They told Thomas that Hill had accused him of sexual harassment. "You've got to be kidding," he said, calling the allegation a "terrible, terrible joke." Thomas denied every one of Hill's charges. He said he was dating other women at the time Hill worked for him. He explained how he helped her get a job at Oral Roberts. He said Hill had remained in contact with him after she left the EEOC. He said he preached against sexual harassment at the EEOC and had never dated anyone on his own staff.[11]

Clifford Faddis, a friend of Thomas's from Missouri, was walking up the driveway to Thomas's house as the FBI agents left. Thomas's face was ashen. "I'm lower than a hog's belly," Thomas told his friend. He seemed more troubled by Hill's betrayal than by the substance of her allegations. "He just kept saying over and over again, 'This was a friend. This was someone I tried to help,'" Faddis recalled.

Later that afternoon the male FBI agent who had interviewed Thomas called and told him the agency viewed the allegations as a "he said/she said" dispute that could not be conclusively resolved.[12] Thomas relaxed a bit. He drove with his friend down to the banks of the Potomac. They sat along the water and reminisced about their days together in Missouri. At the time Faddis was struggling financially. Thomas sought to buck him up, later giving him a copy of Rabbi Harold Kushner's *When Bad Things Happen to Good People.*

The FBI delivered copies of its interviews to the Senate Judiciary Committee and the White House Wednesday afternoon, two days away from a scheduled committee vote on Thomas's nomination. Biden briefed Senate majority leader George Mitchell of Maine, minority leader Robert Dole of Kansas, and Strom Thurmond. "Gentleman, here's our situation," he said. "And I laid out the facts, that she didn't want to go public." Mitchell immediately saw the seriousness of the charges, Biden told Specter, zeroing in on Hoerchner's corroboration of Hill's charges.[13]

Over the next twenty-four hours Biden informed every Democrat on the committee of Hill's charges and Thomas's denials, looking for a signal that he should take further action. Under committee rules, any member could request a delay in the committee's vote. None did.

Meanwhile, Thurmond, the ranking Republican member, discussed Hill's allegations only with Dole. Specter learned of the allegations by chance Thursday evening from Democratic senator Dennis DeConcini of Arizona.

Specter immediately sought out Biden, catching him in the Senate dining room. Biden called for a staffer to give Specter the Hill file. Finding a quiet corner off the Senate floor, Specter began reading. Compared with the allegations that would come later, Hill's allegations seemed "mild" to Specter. Nonetheless, he was concerned.[14] Early the next morning he told Danforth that he wanted to talk to Thomas before the Judiciary Committee voted.

Hill, meanwhile, was angered to learn that her statement had not been circulated to all the committee members. She felt the committee staff was giving her the "runaround."[15] She called two friends in California for advice, and one arranged for her to talk to Senator Paul Simon of Illinois. Hill asked Simon if her statement could be distributed in confidence to all one hundred senators. Simon told her she might as well hand it out to every news organization in the country. After talking to Hill, Simon read the FBI report. Hill's two friends also called Charles Ogletree, a Harvard law professor.[16]

On Friday Ogletree phoned his Harvard colleague Laurence Tribe, author of *God Save This Honorable Court* and one of the leading advocates for aggressive Senate scrutiny of judicial nominations. Tribe had a direct pipeline to the Judiciary Committee; a former student, Ron Klain, was Biden's chief of staff. Tribe called Klain and told him that "a group of women professors from the West Coast" were concerned that the Judiciary Committee wasn't seriously considering a charge of sexual harassment against Thomas. After the call, Klain received Biden's permission to copy Hill's statement and give it to each Democrat on the committee before the scheduled committee vote at ten o'clock that morning.[17]

Before the vote Thomas went to see Specter. He denied Hill's charges categorically. "It just didn't happen," Thomas told Specter. Thomas also told Specter that the allegations conjured the worst stereotypes of black men: predators with large sexual organs who boasted of their prowess. "I wouldn't do that," Thomas said. "Black men are always accused of that."[18] Satisfied, Specter resolved that he would vote to confirm Thomas in committee.

Minutes later Biden strode into the Senate chamber and asked to be recognized. With "heavy heart," Biden announced he would vote against Thomas. The night before, Senator Heflin of Alabama had also announced his intention to vote no, meaning that Thomas could now do no better than a 7 to 7 tie in the fourteen-member committee.

Biden hurried from the Senate floor to the building where the Judiciary Committee was scheduled to vote. In the middle of the hearings

Biden excused himself to call Thomas at home. He told Thomas his no vote had nothing to do with Hill's charges. According to Thomas, Biden told him that if he was attacked over Hill's allegations, he would be Thomas's "biggest defender." (Biden later disputed this characterization of the conversation to Danforth, saying that he would only defend Thomas against rumors of Hill's allegations. If Hill went public, Biden said he would describe his vote against Thomas as opposition to Thomas's philosophy, not his character.)[19]

Just prior to the committee's official vote, Biden said to his colleagues: "I also indicated in my discussion with Judge Thomas that I believe there are certain things that are not at issue at all. And that is his character, or characterization of his character.... This is about what he believes, not about who he is. And I know my colleagues will refrain, and I urge everyone else to refrain, from personalizing this battle."[20]

As expected, the committee split 7 to 7, with only Senator DeConcini crossing over to support Thomas from the Democratic side.

At the White House McClure projected he had close to sixty votes for confirmation. News reports were also writing about the confirmation as a fait accompli. The vote was scheduled for October 8.

To those who knew of Hill's allegations, the committee's vote September 27 signaled that any further investigation of the charges was over. The issue now boiled down to Hill. Would she go public? On Wednesday, October 2, Nan Aron of the Alliance for Justice called Sonia Jarvis, a friend of Hill's, and put the question to her: Was Hill prepared to go public with her allegations? Jarvis called Hill, who said no.

Word of Hill's allegations was now spreading rapidly among all the major groups opposed to Thomas. The weekend after the committee vote they were discussed at two different Washington dinner parties.[21]

Two Washington reporters were also hot on the trail. *Newsday* reporter Timothy Phelps was beginning to connect the dots and on

Wednesday, October 2, quizzed Senator Simon about Hill. By Friday, he was trying to reach Hill in Oklahoma. Nina Totenberg, a national-affairs reporter for National Public Radio, had already tracked her down. Hill was undecided about what to do. At a friend's suggestion, she called Charles Ogletree at Harvard. He suggested she tell Totenberg that she would not talk unless Totenberg proved she had a copy of her statement.

Hill wrestled with the implications of taking her story public. On Friday one of her friends called Aron, Aron said. Would Hill going public be enough to defeat Thomas's nomination? she asked. Aron said no. But if Hill was looking for a national stage to publicize sexual harassment, that was a different story, Aron said. "It will change the world," she said. "It will shine the light where it hasn't been shone."

On Saturday morning, October 5, Hill said she told Totenberg she had no comment unless Totenberg had her statement. Four hours later Totenberg called back and began reading Hill's statement over the telephone. They spoke for forty minutes. Two hours later Totenberg called the White House asking for comment on a story she planned to air the following morning.[22] Phelps, meanwhile, was feverishly preparing a story on Hill for Sunday's edition of *Newsday*, based on reporting that did not come from Hill but was corroborated by her.

Thomas learned the story was about to break in a telephone call from the White House counsel's office that night. "It's the scum story," he told his wife, referring to the name they had given Hill's charges.[23] Thomas sat down and wrote out a list of women who had worked for him who could vouch for his integrity and gave it to the White House press office.

Thomas was convinced the story had been leaked by liberal interest groups. "How can we let these liberal groups do this to me?" he told Duberstein, according to Danforth. "Anita Hill is lying. Can't we just say that?"[24]

That night Thomas also called his family, telling them a story in the news would cause them great personal anguish. "He said he was sorry for the embarrassment and pain it would cause me," recalled Nelson Ambush, Thomas's ex-father-in-law.

Totenberg's story played across the nation Sunday morning, two days before the scheduled Senate vote. Within minutes television camera crews and reporters were staking out Thomas's home in suburban Virginia. Others were catching planes to Norman, Oklahoma, to snag interviews with a previously unknown college professor named Anita Hill. The story dominated network television and radio news.

Totenberg's story on NPR accurately characterized Hill's allegations, but it contained a significant error. The news report said Thomas had told the FBI that he asked Hill out for a date but denied any other inappropriate conduct. In fact, Thomas denied to the FBI that he had ever tried to date Hill.[25] Totenberg quoted an unnamed source—Hoerchner—corroborating Hill's story. But she did not quote from the statements of the two witnesses who told the FBI they never saw inappropriate behavior, because they were contained in the FBI's report that Totenberg did not have. Quoting Senator Paul Simon saying he hadn't seen Hill's statement or the FBI's report, the story also left the impression that Democrats hadn't been aware of the allegations at the time they voted on Thomas in committee. Simon later said he simply had been wrong about telling Totenberg he hadn't seen Hill's statement.

In suburban Virginia Thomas was a prisoner in his home. Television trucks were parked outside, waiting to capture images of the embattled nominee. Inside, Thomas was slipping into a period of deep anguish. He couldn't eat. He couldn't sleep. "He kept saying, 'Why are they trying to destroy me?'" Virginia Thomas told Danforth.[26] Thomas wanted to issue a strong denial, but Duberstein wanted to hold off, believing that would only feed the story. The reality, however, was that the story was already dominating the airwaves.

Hill jumped headlong into the fire Monday afternoon when she called a news conference to discuss her allegations publicly for the first time. "That's it!" Thomas shouted when he heard about it. "They can have it! I give up!"[27]

In the Senate Thomas's support began to crack when Democratic senator Jim Exon of Nebraska, who had said he would vote for Thomas, requested a delay in Tuesday's vote. Danforth now also had doubts. He asked Thomas what he thought. Thomas was adamant: No delay. That night Clarence and Virginia Thomas slept at the Silbermans' Georgetown town house, trying to escape the media.

By Tuesday, the Capitol was bedlam. Hill's Monday news conference was replaying continuously on television. The Capitol switchboard jammed with phone calls to Senate offices. Hill's charges hit a raw nerve. "For American women, the average American woman outside the Beltway, it was like putting your finger in a light socket," said Patricia Schroeder, the Colorado Democrat. That morning Specter had also joined in a call to postpone the confirmation vote.

On the other side of the Capitol, Schroeder and about seven other Democratic women in Congress, believing the Senate was proceeding with a vote, marched over to the Senate and found the Democrats huddled in a conference room. Schroeder banged on the door and demanded the group be admitted but was barred from entering. "It was chaos," she said.

Dole, Danforth, Hatch, and other top Republicans, meanwhile, were huddled in Dole's office considering the options. Fred McClure argued against a delay, saying it would only give Thomas's opponents more time to break down his support. Throughout the meeting, however, the participants were learning of more senators, all previously supportive of Thomas, calling instead for a delay. Dole counted only forty-one solid votes for Thomas, with as many as eighteen undecided. Proceed with a vote, and Thomas was sure to lose. Danforth laid out the situation to Thomas in a phone call, telling Thomas the decision to proceed was his.

Thomas wanted to go forward, even if it meant defeat. He had invested all his energy in anticipating a vote Tuesday. A delay simply meant more agony. He told his wife and the Silbermans he was done. Judge Silberman urged Thomas to tough it out. So did Orrin Hatch and Senator Alan Simpson of Wyoming. Some argued that fighting on was the only way for him to clear his name. Others reminded him of how many people, including the president, believed in him. Ending it now would let them down, too.

Thomas stepped outside into the Silbermans' backyard, a brick-walled enclosure with a small swimming pool. He paced back and forth and around the pool. He lit a big cigar and took deep drags as he wrestled over what to do. The most pivotal moment in the whole ordeal had finally come: Give up or go on. The decision was his. Thinking about that moment years later, Thomas struggled to reconstruct what was going through his mind. "I didn't have any energy to go back and keep fighting," he said. "They'd whipped me down all summer." In the end Thomas said he simply did what everybody was telling him to do.

He called Danforth back with an answer: Delay the vote.

Danforth walked onto the Senate floor and, trembling with anger, asked to be recognized. In fifteen years in the Senate, he had never witnessed such a vicious campaign. "This whole confirmation process has been turned into the worst kind of sleazy political campaign, with no effort spared to assassinate the character of Clarence Thomas," Danforth said, fuming. He assailed his colleagues for calling a delay. "We need a delay," he said, his voice now dripping with sarcasm. "We need more time for the People for the American Way to make their phone calls digging up the dirt. We need the interest groups to have more time for the sharks to gather around the body of Clarence Thomas. Oh, we need a delay."

Fresh from his telephone call with Thomas, Danforth quoted his friend to the chamber: "They have taken from me what I have worked forty-three years to create. They have taken from me what I have taken forty-three years to build—my reputation. I want to clear my name."[28]

Later that night Democratic and Republican leaders struck a deal: They would postpone the vote for one week. In the interim, the Senate Judiciary Committee would reopen hearings on the nomination. They would ask Hill and Thomas to testify. The hearings would begin on Friday, less than three days away.

Having made the decision to fight on, Thomas now discovered he simply had no strength. He hadn't really slept in days. He twitched uncontrollably. He had difficulty breathing. The muscles in his neck tensed like rusty coils.

He asked his wife to call friends to come pray with them. They stood in a circle, holding hands, praying for God's strength. Prayer was the only thing that seemed to help, Virginia Thomas recalled.

President Bush invited Thomas to the White House Wednesday afternoon to show his support. Putting his arm around his beleaguered nominee, Bush walked with him on the South Lawn grounds in full view of the media. "Maybe it would be easier for you if I didn't persevere," Thomas told Bush.

"No, it would make it worse," Bush said. "You deserve to be on the Supreme Court, and we will back you."[29]

The next morning Thomas went to his office to begin preparing for the hearings the next day. Mike Luttig, the Justice Department lawyer who had led Thomas's preparation over the summer, arrived shortly after nine o'clock and found Thomas in a conference room. Thomas looked at him and burst into tears. He bellowed in uncontrollable sobs, hyperventilating as he staggered around the room. "They have ruined me!" Thomas wailed. "I have nothing left anymore."[30]

That afternoon seventeen women who had worked with Thomas over the previous ten years held a news conference. Nine of them, some in tears, went to the microphone and said Thomas was incapable of the conduct Hill described. Their testimonials were all but buried in the news about the hearings set to begin the next day.

Behind closed doors Senate leaders were working out the terms of the upcoming hearings. In a nod to Thomas, Biden gave the nominee his choice of appearing before or after Hill. Thomas chose to go first. Working with Danforth and Dole, Hatch proposed that Republicans delegate the job of questioning the two witnesses. Hatch volunteered to question Thomas. For Hill, Hatch suggested Specter, a skilled prosecutor with years of courtroom experience. Others in the room rejected the idea, according to Hatch. Specter was the most liberal Republican on the committee. He was pro-choice. He might very well vote against Thomas in the end. But Hatch reasoned that making Specter part of the Republican team would make it more difficult for him to ultimately vote against Thomas.

Everyone involved had less than two days to prepare. In a courtroom setting the sides would have had months of preparation, with ability to interview witnesses, take depositions, and gather evidence. Indeed, in a courtroom the allegations would never have been heard because the statute of limitations for filing sexual harassment charges was six months.

The night before the hearing Luttig called Thomas at home to find out how he was doing with his statement. Thomas read some of it over the telephone. It was incoherent. The hearings were twelve hours away, and Thomas didn't have a statement.[31]

Virginia Thomas called a neighbor to come give Thomas a haircut and a head massage. Afterward, he told his wife he was exhausted. He lay down to sleep. He tossed and turned fitfully before getting up at 1:30 A.M. He was due before the Senate in less than nine hours. He sat down at the kitchen table in front of a yellow legal pad. His thoughts were suddenly clear. Faced with a deadline, he wrote quickly, handing completed pages to Virginia to type on the word processor. Shortly before five o'clock, Thomas set down his pen and went to bed. An hour later, Kay and Charles James arrived to pray before Thomas left for his appointment with the American people.

. . .

Shortly before ten o'clock Thomas, Danforth, and their wives squeezed into Danforth's private Senate bathroom, the only place where Danforth figured the four could be alone. Danforth pushed the play button on a tape recorder brought in for the occasion. Holding hands in a tight circle, the two couples joined the music and belted out the words to "Onward, Christian Soldiers." When the music ended, Danforth clasped both hands on Thomas's shoulders. The Episcopalian priest looked Thomas in the eye and said, "Go forth in the name of Christ, trusting in the power of the Holy Spirit."

The music was a call to battle; indeed, for Clarence Thomas and his supporters the nomination had become a holy war. Over the previous forty-eight hours Thomas had come to believe that the battle was less about himself and more about American decency, and the ethics of using political differences to justify personal attacks.

"In my forty-three years on this earth," Thomas said, moments later at the witness table, "I have been able, with the help of others and with the help of God, to defy poverty, avoid prison, overcome segregation, bigotry, racism, and obtain one of the finest educations available in this country. But I have not been able to overcome this process. This is worse than any obstacle or anything that I have ever faced.

"No job is worth what I have been through, no job. No horror in my life has been so debilitating. Confirm me if you want, don't confirm me if you are so led, but let this process end."

Thomas was almost finished. Yet he still hadn't made it clear whether he was withdrawing or fighting on. "I will not provide the rope for my own lynching or for further humiliation," he said in conclusion. He stopped reading. Thomas was not withdrawing.

Hill appeared confident and poised as she took her place at the witness table. Arriving in Washington days earlier, she had been closeted with a dozen lawyers and advisors, all helping her prepare. The team included Harvard's Ogletree and noted Supreme Court lawyer John Frank, who had argued the historic *Miranda* case. A Democratic political consultant also attended some of the prep sessions for Hill, along with several other female law professors who had expertise in sexual harassment.

Hill's testimony reiterated the allegations against Thomas made in her written statement. Midway through her testimony, however, Hill electrified the hearing room with a new wrinkle.

"One of the oddest episodes I remember was an occasion in which Thomas was drinking a Coke in his office, he got up from the table, at which we were working, went over to his desk to get the Coke, looked at the can and asked, 'Who has put pubic hair on my Coke?'" she said calmly. "On other occasions he referred to the size of his own penis as being larger than normal and he also spoke on some occasions of the pleasures he had given to women with oral sex."

Hill said it was only after being approached by Democratic staff in the Senate that she resolved to come forward. "I felt that I had a duty to report," she said. "I could not keep silent."

Hill also sought to explain her telephone calls to Thomas after she had left the EEOC. Thomas had produced phone logs documenting eleven phone messages from Hill. Some included cordial greetings; others were requests for advice.[32] Thomas's supporters had been waving copies of the phone logs for three days, attempting to show that Hill's behavior was inconsistent with her allegations. "I did not want, early on, to burn all the bridges to the EEOC," she testified. "Perhaps I should have taken angry or even militant steps, both when I was in the agency or after I had left it, but I must confess to the world that the course that I took seemed the better, as well as the easier approach."[33]

Virginia Thomas watched Hill testify from home. She believed as many did that Hill was credible. But she could not understand her behavior in light of her own experience. If what she said was true, why hadn't she done something?

Hill remained on the witness table throughout the day. More salacious details followed. Hill said Thomas spoke frequently of a pornographic movie character named Long John Silver, an apparent reference to a character actually named Long Dong Silver. She said Thomas frequently described the length of his penis in reference to the movie character. Other times he discussed the size of women's breasts, she said.

The charges made Thomas's behavior appear all the more damning. Yet they also raised the stakes for Hill. The more disgusting Thomas's behavior appeared, the more incredible it seemed that Hill put up with it, much less agreed to follow Thomas to the EEOC. First Lady Barbara Bush jotted some notes to herself in her diary that day. "I do not mean to sit in judgment, but I will never believe that she, a Yale Law School graduate, a woman of the 80s, would put up with harassment for one moment, much less follow the harasser from job to job."[34]

Ironically, Hill's charges helped, not hurt Thomas among many blacks, according to Barbara Arnwine, who was then executive director of the Lawyers' Committee for Civil Rights Under Law. "Here was another black woman trying to destroy a black man," said Arnwine. "People saw her as a tool of the feminists, that she was being used to stop him from getting this position."

Hill's new charges troubled Specter because they hadn't been included in her original statements. To him it seemed as though Hill had embellished and embroidered her testimony, and he quizzed her sharply about the discrepancies. Hill said she had been uncomfortable discussing such graphic, personal information.

Later, Specter believed Hill perjured herself over the characterizations of her contacts with committee staff leading up to her decision to submit a statement. *USA Today* reported that Hill was told her statement would be shown to Thomas, and that he would then withdraw rather than expose himself to public humiliation. What did Hill know about that? Specter asked. The point was crucial, Specter thought, because it showed that Hill might have been motivated by a desire to torpedo Thomas's nomination with minimum exposure to herself. Hill at first denied that Senate staffers ever suggested such a scenario, but Specter pressed her on it.

Over a break, Biden ordered staff members to tell Hill's lawyers that their client's memory needed to be refreshed about the phone contacts by the time she returned to testify. Biden also was convinced Hill was lying, he later told Specter.[35]

Later that afternoon Hill testified that her discussions with Senate Democrats had included "something to the effect that the information might be presented to the candidate and to the White House. There was some indication that the candidate, excuse me, the nominee might not wish to continue the process."[36] Although his interpretation was sharply disputed by others, including Hill, Specter believed he had cornered Hill in "flat-out perjury."[37]

Shortly before eight o'clock Hill finished her testimony and left the Senate. Thomas paced Danforth's office nervously, recalled Kay James, who waited with him. He hadn't watched Hill testify, relying on his wife to give him updates throughout the day. He felt certain that the public believed her. He was just a big black man, preying on defenseless women. From the South, Thomas knew the stereotype. He knew what people, especially white people, thought in the deepest recesses of their minds. Who would believe him?

Sensing Thomas's anxiety, James picked up the phone and called her brother, Lucky, who had watched Hill testify. "Lucky," James told her brother, "tell Judge Thomas what you think he ought to say." James handed the phone to Thomas. Lucky shouted into the receiver: "Get out there and don't take no shit! Let 'em know they been in a fight, man. I'm with you. The people are with you!"

It was the last voice Clarence Thomas heard before he walked back out to face the television cameras.

Thomas took his seat at the witness table for the second time that day. He had no notes or written remarks. He sat waiting to speak, smoldering in anger. He cut Biden a look of utter disgust as the chairman acknowledged the "tough day, tough night" of the previous twelve hours. When Biden finally nodded for him to speak, Thomas sat bolt upright and leaned over the witness table, looking as though he might suddenly stand up and knock the table over.

He categorically denied each of Hill's charges.

"Today is a travesty," Thomas said. "I think that it is disgusting. I think that this hearing should never occur in America. This is a case in which this sleaze, this dirt was searched for by staffers of members of this committee, was then leaked to the media, and this committee and this body validated it and displayed it in prime time over our entire nation.

"The Supreme Court is not worth it. No job is worth it. I am not here for that. I am here for my name, my family, my life, and my integrity."

Thomas sat back in his chair momentarily before beginning to speak again. "And from my standpoint, as a black American, as far as I am concerned, it is a high-tech lynching for uppity blacks who in any way deign to think for themselves, to do for themselves, to have different ideas, and it is a message that, unless you kowtow to an old order, this is what will happen to you. You will be lynched, destroyed, caricatured by a committee of the U.S. Senate, rather than hung from a tree."[38]

Utter silence hung over the committee room. Thomas had just spoken what was to become one of the most memorable lines in American judicial history.

The comment electrified many of his supporters, but some, including Thomas's closest friends, thought it out of character for a man who was devoted to minimizing racial differences. Thomas had scored civil rights leaders for playing the "race card." Yet with his back to the wall, he threw it down on the table before millions.

Years later Thomas said he felt his lynching analogy was appropriate because charges of sexual misconduct were the only ones a black man could never defend against. The stereotype of the sexually rapacious black man, constructed from whole cloth out of the white South's fear of "race mixing," was too ingrained in the American psyche.

"I was ticked. I was unhappy," he said. "Somebody says something like this against you, what are you going to do? You are going to sit there and be perfectly calm about it?"

Thomas amplified his racially charged comments when he returned to testify the next day. "Language about the sexual prowess of black men, language about the sex organs of black men, and the sizes, et cetera, that kind of language has been used about black men as long as I have been on the face of this earth," he testified. "I cannot shake off these accusations because they play to the worst stereotypes we have about black men in this country."

While Thomas was testifying, a former EEOC employee named Angela Wright had come forward to accuse Thomas of making inappropriate comments to her six years earlier. Wright told Senate investigators the comments didn't constitute sexual harassment, nor was she intimidated by Thomas. "Annoying and obnoxious" was how she described Thomas's behavior.[39] Her testimony could lend credibility to Hill, but Wright had been fired by Thomas, making her motives potentially suspect. In the end, Democrats and Republicans agreed not to call her to testify but to make her deposition part of the record.

Thomas finished testifying Saturday evening. Tired and depleted, he dined that night at Morton's steakhouse in suburban Virginia and happened to run into Robert Bork. The two men embraced as the restaurant crowd broke into applause, said Orrin Hatch, who was there.

All day Sunday the committee took testimony from witnesses corroborating either Hill or Thomas. Two witnesses said the law professor in the early 1980s had discussed sexual harassment by Thomas, and two said she referred to an unnamed supervisor who had harassed her. None of the witnesses worked with Hill at either the Department of Education or EEOC, and all were attempting to recall conversations up to a decade old.

Thomas's witnesses included three women who worked with both Hill and Thomas. All three said they never witnessed any inappropriate behavior. They also painted a starkly different portrait of Hill. "On Friday, she played the role of a meek, innocent, shy Baptist girl from the South who was a victim of this big, bad man," said J. C. Alverez. "I don't know who she is trying to kid. Because the Anita Hill that I knew and worked with was nothing like that. She was a very hard, tough woman. She was opinionated. She was arrogant. She was a relentless debater. And she was the kind of woman who always made you feel like she was not going to be messed with, like she was not going to take anything from anyone."[40]

The testimony prompted another EEOC staffer to write the committee. "If you were young, black, female, and reasonably attractive, you knew full well you were being inspected and auditioned as a female," wrote Sukari Hardnett, who worked at the EEOC in 1985 and 1986. "Women know when there are sexual dimensions to the attention they are receiving. And there was never any doubt about that dimension in Clarence Thomas's office."[41]

The last and final panel of witnesses, eight former female colleagues of Thomas's, took their places before the committee after midnight Sunday. Thomas's former chief of staff, Pam Talkin, said he never so much as swore in her presence. "I have never worked in a work environment where any individual, man or woman, was more committed to establishing a workplace free from discrimination and harassment," she said as the hearing drew to a close. "It is the saddest of ironies to me that the behavior that Judge Thomas found most abhorrent is the behavior that he is now being accused of."

The contradictions between the testimony of Hill and her supporters and Thomas and his were never resolved. No one who testified overheard or witnessed Thomas making crude remarks to Hill or asking her out on dates, and their former colleagues at the Department of Education and the EEOC backed Thomas, not Hill. As Thomas's male friends knew, Thomas could be coarse and irreverent among friends and in social settings. But he was also an extremely careful man, and he drew a sharp line between business and pleasure. That he might repeatedly menace a subordinate with aggressive sexual language simply seemed unbelievable to those who knew him well and had worked with him for long periods.

A fact glossed over during the hearings was that Thomas and Hill knew each other socially before they knew each other professionally. The two had met through Gil Hardy in 1981, when Thomas was just thirty-two, separated from Kathy, and living on his best friend's sofa in the same apartment building as Hill. All three lawyers were part of a small, close-knit circle of black Yale Law School graduates living and working in Washington. Thomas and Hill also both alluded to a casual, social relationship while they worked together. Thomas sometimes gave Hill a ride home from work, and on a few occasions Hill invited him in for a drink. Neither Thomas nor Hill talked much about their relationship outside the office, and Hardy's death prevented any clarification from him. (Thomas told friends he believed Hill would never have come forward if Hardy had been alive.) But the fact that they had a personal relationship suggested other possible explanations for the context of Thomas's alleged remarks. Understanding that context, so crucial to validating a charge of sexual harassment, would have been difficult even in the best of circumstances, no less ten years later. But it became impossible in the excessively partisan and public spectacle of the Senate hearings.

In the end Thomas and Hill remained the only two people who knew what transpired between them, and each told a different story. Although it was plausible that Thomas said what Hill alleged, it seemed implausible that he said it all in the manner Hill described.

Bullying a woman simply wasn't in Thomas's nature and ran contrary to how he conducted himself around others in a professional environment. And if the context wasn't as Hill alleged, was it fair to turn private conduct into a political weapon to defeat his nomination?

Members of the Senate had less than forty-eight hours to decide.

By previous agreement the Senate was scheduled to vote on Thomas's nomination at 6 P.M. Tuesday. Thomas's opponents calculated fewer than eight undecided votes and about ten senators whose support might be soft enough to peel away. The coalition lobbied not only the senators but also their wives.

All the senators found something in the record of the hearings to justify their position. Democrats in particular seemed angry over Thomas's "high-tech lynching" comment. "He used his race in a way that he had always refused to do," said Democratic senator Bill Bradley of New Jersey. Robert Byrd of West Virginia, a conservative Democrat, said he was prepared to vote for Thomas until he claimed racism. "It was a diversionary tactic," the onetime Klansman said angrily. "I thought we were past that stage."

The outcome was so much in doubt that Vice President Dan Quayle presided over the Senate in case he needed to break a tie vote. By lunchtime, fifty senators had announced in favor of Thomas, but the figure included two whose votes were suddenly in doubt. Dole informed Danforth that Senators William Cohen of Maine and Warren Rudman of New Hampshire were reconsidering. Without them, Danforth believed the vote for Thomas was lost.

An hour before the vote, Danforth took the Senate floor one last time. The three and a half months of the confirmation ordeal—lobbying his colleagues, holding news conferences, denying Hill's charges to the Senate—had taken its toll on the Missouri Republican. "There is no joy in these proceedings," Danforth said softly, his voice limp with exhaustion. "The joy that we experienced three and a half months ago has turned to pain, and the best that can be said is that in approximately another hour, there will be a feeling of relief at the determination one way or another."[42] The clerk called the roll.

Six hundred miles away, Thomas's childhood friends gathered around a big-screen TV in Pin Point outside the mobile home of Abraham Famble, Thomas's boyhood playmate. CNN wired a satellite feed of the Senate vote to Thomas's birthplace and sent a producer down to capture the scene. Scores of people, many of whom had never even met Thomas, crowded on Famble's lawn, sitting on picnic tables and watching the votes being cast. The network kept a running tally of the votes. The number for confirmation was creeping into the high forties. The excitement of the crowd was building before a technical foul-up suddenly severed the feed.

In the Senate, Thurmond, the onetime segregationist from South Carolina, cast the fiftieth vote to confirm Thomas to the Supreme Court. Bob Packwood of Oregon and Jim Jeffords of Vermont broke Republican ranks and voted against Thomas, joining forty-six Democrats. Rudman and Cohen stayed in the fold. In the end Thomas prevailed, 52 to 48, the highest number of negative votes for any Supreme Court justice in American history. Without Danforth's last-minute lobbying, and the prestige of his fourteen years in the Senate, Thomas probably would have lost. Rudman and Cohen told Danforth afterward that they voted for Thomas only out of loyalty to him.

Back in Pin Point, CNN restored the satellite feed just after the final votes were tallied. The crowd erupted in a cacophony of whoops and hollers. It was too much to believe, Famble thought. His friend, a barefoot kid from Pin Point, Georgia, was a justice of the Supreme Court. The journey had taken Thomas all of forty-three years.

Clifford Faddis reached Thomas's home in Alexandria, Virginia, just as the voting ended. Thomas, still wet from a shower, grinned broadly as Faddis gave him a big bear hug. Two other couples were already inside. Pausing for a moment, the group formed a circle in the living room. They held hands and prayed, thanking God for Thomas's victory. Danforth, looking drained and exhausted, arrived, followed shortly by Senators Thurmond and Simpson and lawyers from the White House counsel's office. It was a tight squeeze inside Thomas's small living room. Someone ordered Domino's pizza, and the delivery boy turned out to be the son of Trent Lott, Republican senator from Mississippi. Over beer and wine, they began to loosen up after the monthslong ordeal.

Outside, the news media waited in the rain for Thomas to make a statement. As he prepared to go outside and say a few words, U.S. marshals handed him a bulletproof vest. Agents told the newest Supreme Court justice that they thought his life was in danger. Thomas put the vest on underneath his suit jacket. The bulky contours of the armor made him appear overweight.

Outside, Thomas credited God for his victory. "I think that no matter how difficult or how painful the process has been, that this is a time for healing in our country, that we have to put these things behind us," he said.

The nation, as it turned out, was not so easily healed. Neither was Clarence Thomas.

CHAPTER 15

WOUNDED BEAR

A brilliant blue sky stretched overhead when Thomas, joined
by hundreds of friends and supporters, raised his hand on
October 15 to take the constitutional oath of office on the
White House lawn.

In his remarks President Bush barely acknowledged the tumult of
the previous weeks. "America is blessed to have a man of this character serve on its highest court," said Bush. The crowd erupted in rau-
cous applause after Thomas completed the oath.

"There have been many difficult days as we all went through the
confirmation battle," Thomas said afterward. "But on this sunny day
in October, at the White House, there is joy. Joy in the morning." In a
nod to his wife, Thomas concluded with her words: "Only in America
could this day have been possible."

Inside the White House, away from the cameras, Thomas hugged
his family and friends, bouncing from one group to the next. Pictures
from the event show him smiling broadly, arms draped around the
shoulders of his friends and beaming next to his long-lost father,
M. C. Thomas, who had arrived from Philadelphia. The most notable
absence, perhaps, was Myers Anderson, who had hoped his grandson
might one day have a "coat-and-tie job."

It was a bittersweet moment. Thomas had won confirmation at the cost of his reputation. Physically exhausted and emotionally drained, he hadn't really had time to examine the depth of his wounds or think about the future. A former classmate later wrote that Thomas's triumph reminded him of a mother who gave birth to twins, then learned one of the babies was dead. Should he offer congratulations or condolences? "Yeah, that's what it was like," Thomas told him.

Some media commentators, meanwhile, had already delivered their postmortems: Thomas was damaged goods, destined to be a Supreme Court footnote before his tenure even began.

Five days later Chief Justice William H. Rehnquist administered the required Supreme Court oath to Thomas, officially making him the nation's 106th Supreme Court justice, in a closed ceremony attended by his wife and Senator Danforth. A more public ceremony had been scheduled for the following week, but Thomas wanted to start working—to begin putting the past behind him.

At forty-three, Thomas was half the age of some of his older brethren. Justice Byron White, seventy-four, had been named to the Court when Thomas was entering high school. Justice Harry Blackmun was eighty-two, old enough to be Thomas's grandfather. Thomas was still paying off his student loans. Thomas believed that his youth, which had been so attractive to those who advanced him for the judiciary, was yet another liability. He'd wanted more time at the court of appeals to gain experience. The age and experience of his new colleagues intimidated him and added to the already crushing weight he felt to prove himself.

The other justices welcomed Thomas graciously and encouraged him to seek their help. He learned, however, that the offers were mostly perfunctory in an institution where independence reigns. The justices hear and then vote on cases as a group, but each functions autonomously. They are coequals, as insulated from one another as they are from the pressures of the media, the legal community, and presidents who appoint them. Confirmed for life, they are beholden to no one, following only their own conscience and their own reading of the law. Thomas was on his own and he knew it. As Rehnquist wryly told him, Thomas would spend his first five years wondering how he got to the Supreme Court and the rest of the time figuring out how the rest of his colleagues got there.

No one was freer with his advice, perhaps, than Thurgood Marshall, who talked to Thomas for more than two hours when Thomas paid him a courtesy call.

Thomas and Marshall came from different eras, and they were different people. Marshall was gruff and cranky. He owed no man deference, showing his irascibility with occasional angry outbursts. In his own departing news conference, he told the world he saw no difference between a white or a black snake, an apparent reference to Thomas as his likely successor. What galled Marshall more than anything was the notion that he could be replaced by anyone, much less a conservative black man like Thomas. "They think he's as good as I am," Marshall grumbled about Thomas, according to a former clerk quoted in his biography.[1]

Marshall believed America and its great Constitution never were and never could be color-blind. He devoted his life to writing color into the law, using America's historic discrimination of black Americans to make race a factor in public-school education, college admissions, and the workplace. By contrast, Thomas believed America could only become color-blind by expunging color from the law. Marshall spent a lifetime reminding America of its racially stained past; Thomas planned to spend a lifetime eliminating race from American law. Marshall's era was over, and Thomas's was about to begin.

One pearl of wisdom from Marshall stuck with Thomas. As a black man, Marshall said, Thomas would be held to a different standard. The advice resonated instantly with Thomas because his grandfather also had preached that a black man in America had to work twice as hard to get half as far. All the questions about his qualifications had reinforced that lesson.

Thomas learned later that he would be held to a different standard by black Americans as well as white. And that standard was the great Marshall himself.

Facing the U.S. Capitol the Supreme Court looks like a cross between Fort Knox and the New York Public Library. The walls are hewn of thick, white marble. The weight of the exterior stone makes the building appear anchored to the ground with the force of a skyscraper. Inside, each of the nine justices has a suite, facing the streets outside. The chambers themselves are small by modern standards, but they are as secure as bank vaults. Thick, oak doors seal off each chamber from the inner hallways. The marble floors are covered with heavy fabric runners that muffle passing footsteps. Marshals patrol the hallways, hushing boisterous members of the public. The primary business of justices is reading and writing, and every feature of the Supreme Court building is designed to foster peace and quiet.

On the day of Thomas's oath taking with Rehnquist, Danforth walked through the Court and told Thomas it reminded him of a mausoleum. "Where are all the bodies?" he asked.

In truth, the Court's tomblike surroundings were perfectly constructed for a man who had just told the nation that he had died a thousand deaths. The Court insulated and protected Thomas from the moment he set foot inside.

In his first months at the Court, Thomas relied on the same work ethic and discipline that carried him through Saint John's, Holy Cross, and Yale Law School. He awoke early, often by four o'clock, and arrived at the Court before five-thirty. Most mornings he was the only employee—other than the security guards—in the building. The quiet and solitude appealed to Thomas, and this was the only time he could fully concentrate and think. The workload of the Supreme Court was far heavier than that at the court of appeals. Cases for review arrived in metal carts. Every day brought another set of metal carts. The amount of reading was staggering. Each case could include hundreds of pages of briefs and supporting documents.

The workload was even heavier for Thomas because he'd started more than three weeks after the Court term began. All through the summer and early fall the other justices had been setting aside cases that required the assent of one more justice to be taken on appeal. Under Court rules, the votes of four justices are required to take an appeal; those with three were waiting for Thomas's review.

Thomas needed all his strength but soon found that he had none, so wasted was he from the previous three months' ordeal. He was constantly fatigued and repeatedly told his clerks he couldn't focus or concentrate.

Thomas's own stubbornness was part of the problem. Feeling like he could not afford to lose even an hour, he pushed himself to work all the time. He worked weekends, too, or brought work home with him. Thomas ventured outside little, in part because he felt he had no time. But he also discovered that he was no longer anonymous. The media frenzy of the preceding months made him instantly recognizable, and the exposure made him uncomfortable. At the court of appeals Thomas had routinely walked over to the food court at Washington's Union Station for lunch, but such nonchalant visits were now out of the question.

Thomas's mail signaled his sudden notoriety. He received more than twenty thousand pieces of mail—some of it arriving by the sack—in his first year on the Court. Some letters included Bible verses, others contained gift certificates to McDonald's. Staff eventually responded to all of them, Thomas said, and Thomas personally thanked those who had played active roles in support of his nomination.

Thomas signed a photograph to Kay James's brother, Lucky, the man who had told him to take "no shit" from the Senate Judiciary Committee. "Thank you for your wise counsel and advice during my confirmation process," wrote Thomas, according to James.

Thomas's only diversion—his wife's, as well—consisted of finding a new home. They were eager to move from the fishbowl they had been living in. After relentless months in the spotlight, they wanted privacy and some semblance of the life they led before Thomas became a household name. Now making nearly $155,000 a year, Thomas could also afford something bigger. They scouted the suburbs of Virginia and found a wooded, five-acre lot in Fairfax Station, in a relatively new subdivision with only a handful of possible home sites. The lot backed up to woods. After settling on the plans, they sited the new house so it would be completely obscured from the street. "They wanted to go way out where nobody would find them," said Ricky Silberman.

Thomas's need for isolation was just one sign of how confirmation had changed him. There were other signs, too. He stopped reading newspapers or watching television, except for sports programs. At the EEOC he had consumed five newspapers a day. Now he avoided the news. He also stopped laughing. He smiled and was warm with friends, but the full belly laugh was gone. The contrast worried and alarmed them.

Thomas continued to talk about his confirmation and Anita Hill to friends. His refrain was always the same: Why? Why had she done it? He lashed out at the Senate. He seethed over the double standard with which the liberals, the media, and the Senate had treated his nomination. He felt as though the media put the burden on him to disprove Hill's accusations. He was angry—very, very angry, friends who saw him those first few months recalled. Sometimes he sought solace in the chambers of Justice Antonin Scalia, who became one of his early friends on the Court.

"His first couple of years he was beaten down," Scalia recalled. "It took him several years to get over the beating he took. He was very bitter about it."

Thomas remains bitter about the experience today. "It's like someone faking with a left hook and hitting me with the right. I'd been watching for the obvious bigots, and the people who claimed not to be did to me what I thought the others would have done," he said. "I understand the attacks on other levels, but that one I don't understand."

Thomas found one solace: He was in a position to outlive and outlast those who had sought to bring him down. With a lifetime appointment, he would have the last laugh, he told visitors.

In February 1992 *Reason* magazine published an article on his confirmation that Thomas later said came closest to capturing his feelings. Author Edith Efron presented Thomas's confirmation as a crescendo of racial stereotypes. The strategy of vague answers, so effective for Souter, made Thomas appear dumb, shifty, and evasive. The presence of so many white coaches and handlers then made Thomas appear the docile, "good" black doing the bidding of his white "masters." The most devastating stereotype came last: the predatory black man who bragged about the size of his sexual organs. "The oldest and most murderous racist stereotype directed at the black male," Efron wrote.[2]

"He was made out to be some kind of dirty old man, dirty middle-aged man, based on the say-so of one woman," said Scalia.

Thomas also saw the attacks on his beliefs about self-reliance for black Americans as a trashing of his grandfather's values. It was as though, Thomas thought, his entire life experience had been dismissed. "If you grew up in Pin Point and in Savannah," he explained later, "what would you have to have . . . inside your heart, or your soul or your mind to want to get out? . . . That's what died. And it was probably the rest of my youth or innocence that went with it. And it doesn't mean you can't reclaim some things, like the way I've reclaimed a positive attitude. But you can't get it back."

Virginia Thomas, too, was struggling to overcome the aftermath of the confirmation, and it haunted her wherever she went. She studied the faces of the people she met in public. What were they thinking? Did they believe her husband? What did they think of her? Over time, the questions went away, but she still felt shame that she had not done enough to defend her husband. During the confirmation process she was under orders not to speak publicly. Now she bitterly regretted following the advice. Why hadn't she stood up? Why hadn't she told the world about the man she knew and loved? The questions still plague her today.

There was perhaps one silver lining for both husband and wife: The pain of the confirmation proceedings had deepened their relationship in ways that neither could have imagined. Married barely five years, they already were working together like a couple who had lived together a lifetime. The heat of the ordeal bonded them together like alloy, they often said to each other.

Outside the protective walls of the Court and their relationship, however, the storm continued to rage over Thomas's confirmation. Women's groups vowed retribution at the polls in the 1992 presidential campaign year, casting Thomas's nomination as a wake-up call to every woman voter in America. Women's groups had lost the battle to defeat Thomas, but Thomas handed them the biggest campaign issue of the decade.

At the University of Chicago Orrin Hatch was almost shouted off the stage by angry hecklers screaming "Pig!" "Swine!" and "Fascist!"[3] Facing reelection in Pennsylvania, Arlen Specter ran against a woman who campaigned openly on Specter's role in cross-examining Hill.

Across the country, businesses big and small were ripping up sexual-harassment guidelines and starting over with new ones, or publicizing policies for the first time. Consultants used the Thomas/Hill hearings as teaching aids in a suddenly booming business for sexual-harassment training. The volume of harassment charges shot up at the EEOC, Thomas's old stomping ground, and commentators wrote of a seismic shift in workplace relationships between men and women.

Thomas's first major decision on the Court, a dissent involving the Eighth Amendment's prohibition against cruel and unusual punishment, played a major role in shaping his evolving public image.

Keith Hudson was a Louisiana prison inmate punched in the

mouth, eyes, chest, and stomach by a guard while shackled and hand-cuffed. Hudson suffered what a lower court determined were minor bruises, and swelling to his face and lips. The punches loosened his teeth and cracked a dental plate, rendering it useless for several months. At trial, Hudson was awarded eight hundred dollars for his injuries. The Fifth Circuit Court of Appeals reversed the decision, saying that "minor" injuries weren't enough to justify Hudson's claim of cruel and unusual punishment.

The legal question posed by the case was whether such "minor injuries" could trigger a claim of cruel and unusual punishment under the Constitution. Seven of the court's nine justices said that it did.

Thomas, joined by Scalia, said that it didn't. "Today's expansion of the Cruel and Unusual Punishment Clause beyond all bounds of history and precedent is, I suspect, yet another manifestation of the pervasive view that the Federal Constitution must address all ills in our society," Thomas wrote. "Abusive behavior by prison guards is deplorable conduct that properly evokes outrage and contempt. But that does not mean that it is invariably unconstitutional."[4]

Thomas based his dissent on his reading of the Court's Eighth Amendment precedents and the history of how the restriction on punishment had been applied. Earlier courts had interpreted cruel and unusual punishment to apply only to a felon's sentence, not the prison conditions they were subjected to. In the 1970s and 1980s the Court made several rulings that set standards for prison conditions as well. In *Hudson* the Court now said that minor injuries could also merit a claim if the inmate proved the force was applied "maliciously and sadistically."

Thomas anchored his argument in *Hudson* in a close reading of early American history. He compared the language of the Eighth Amendment against historical records of the time and concluded that the amendment had applied only to prison sentences. Thomas argued that it was wrong to expand the scope of the amendment beyond its original meaning. In Court jargon, this method is called "originalism," and over time Thomas became one of its most faithful adherents on the Court.

Rooted in history and focused on a plain reading of the Constitution, originalism sometimes produces decisions at odds with popular norms of justice. In the *Hudson* case, the majority opinion seemed to more accurately reflect popular sentiment that the Constitution should prevent prisoners from being beaten by their keepers.

The dissenters, especially Thomas, were excoriated. Commentators seized on the detail that Hudson was black, suggesting that Thomas should have been doubly sympathetic to his case. A scathing *New York Times* editorial called Thomas "the youngest, cruelest justice," beginning a trend of media coverage that often portrayed Thomas as heartless and unsympathetic to human suffering.

The furor over Thomas's dissent in *Hudson* overshadowed two other instances where he sided with the Court majority in cases roundly applauded by liberal constituencies. In a case from his home state of Georgia, Thomas joined the Court to prevent two white defendants from using jury challenges to strike black jurors.[5] Thomas also joined a near-unanimous Court in ordering Mississippi to take more steps to eliminate the state's black and white university systems. In his concurring opinion, however, Thomas spelled out that those steps should not dismantle the state's historically black colleges, a particular concern of civil rights organizations. "It would be ironic, to say the least, if the institutions that sustained blacks during segregation were themselves destroyed in an effort to combat its vestiges," Thomas wrote.[6]

The ruling that dominated the publicity over Thomas's first term came down three days after the Mississippi ruling. In a 5 to 4 decision the Court upheld new restrictions on abortion in Pennsylvania but reaffirmed *Roe* v. *Wade. Planned Parenthood of Southeastern Pennsylvania* v. *Casey* provided a direct opportunity for the Court to overrule *Roe*, but Justices O'Connor, Souter, and Kennedy crafted a majority opinion that upheld the landmark case. Chief Justice Rehnquist and Justice

Scalia both wrote stinging dissents, asserting that the Court should have overturned *Roe* and left the legality of abortion to the states. Thomas signed both dissents.

The case was Thomas's first on abortion, and it confirmed the charge from liberal critics that Thomas would vote against *Roe*. Conservatives, however, applauded Thomas.

The abortion ruling also launched the popular fiction that Thomas, far from being an independent thinker, was nothing more than the clone of Justice Scalia. The perception dogged him for nearly a decade. For Thomas the comparisons carried an obvious racist subtext: The Court's only black man relied on a white colleague to write his opinions and tell him how to vote. The presumption was patently absurd. No one had ever told Thomas what to say or how to think.

Scalia also resented the double standard. No one, for example, had ever accused Marshall of being Brennan's double even though the two liberal jurists voted together nine out of every ten times and consulted each other regularly on important decisions. For Scalia, the difference was that he and Thomas were conservatives. For Thomas, the difference was his race.

The irony of the comparisons, perhaps, is that Thomas and Scalia are actually very much alike, despite obvious differences in style. Scalia is the Supreme Court's high inquisitor, a sharp-tongued debater whose aggressive questioning of lawyers intimidates and delights, depending on his mood. Scalia leans over the bench, glowering at the lawyers beneath him from behind a pair of bushy eyebrows, and peppers them with questions. His extroverted style is the very opposite of Thomas's reticence. In private, however, Scalia can be quiet and deferential, while Thomas fills the room with his explosive presence.

Thomas and Scalia share similar backgrounds and temperament. Like Thomas, Scalia was raised in a working-class home, the only son of Italian immigrants to the New Jersey suburbs of New York (later the family moved to Queens). Scalia's father, a native of Sicily, put himself through college in New Jersey to become a teacher of Romance languages, pulling his family into the American middle class with the same work ethic that guided Myers Anderson. Catholicism was central to his childhood, just as it was for Thomas, and Scalia remains a devout Catholic today. Tough and uncompromising, Scalia likes being the "skunk at the garden party," as he has described himself on more than one occasion, and he frequently writes dissents in which he stands alone against the Court.

"I think that's something we very much have in common," says Scalia of Thomas. "He tends to be a contrarian, as I do. I find myself in a large majority and I feel uncomfortable. It questions the premises I'm operating on. I think there is a streak of that in Clarence."

More significant, perhaps, Scalia's jurisprudence incubated in the same conservative environment that Thomas's did. Along with Robert Bork, Scalia was one of President Reagan's first appointees to the U.S. Court of Appeals for the D.C. Circuit. Scalia shared the conservative movement's hostility to the perceived excesses of the Warren and Burger Supreme Courts—just as Thomas did—and embraced a narrower, more literal reading of the Constitution. "He is really the only justice whose basic approach to the law is the same as mine," says Scalia. "Isolation does a lot of things. It produces timidity, among other things . . . boldness comes from company. So I appreciated his being here, and I think he appreciated mine."

The similar philosophies, however, didn't compute in race-conscious Washington. Even black civil rights leaders fostered the misconception that Scalia drafted Thomas's opinions or told him how to vote, a belief that many still hold today. It is also striking that Thomas attracts far more negative publicity than Scalia, even though their opinions are often similar.

The abortion ruling marked the end of the Court's term. The summer break was barely under way when Thomas collapsed, his lungs dangerously full of fluid. The four long months of his confirmation battle and the eight months of the Court's term finally caught up with him. Thomas was sick for two months that summer.

As his strength returned, Thomas threw himself into supervising construction of his new house in Fairfax Station and befriended the construction workers on the site. He also discovered Rush Limbaugh that summer. He listened for hours on end to the conservative commentator blast the liberal media and excoriate Democrats on Capitol Hill. Limbaugh's rants gave voice to his own anger.

Political candidates, meanwhile, were using Thomas as a platform for public office. A record number of women were campaigning, and for many of them the Senate's treatment of Anita Hill became a campaign theme. Hill made headlines in August when she accepted an award at the American Bar Association's annual meeting. She received a thirty-second standing ovation at a crowded luncheon where scalpers were selling tickets at almost twice their face value. The ABA's star treatment of Hill was perhaps the clearest signal that the mood of the country was shifting away from Clarence Thomas.

In October 1992 several large news organizations reported polls showing that more Americans now believed Hill than Thomas. The levels of support for each were practically reversed from the previous year. Then more Americans saw Thomas as the victim of a last-minute attack from an ungrateful employee. Now Hill was the victim, callously scorned in the Senate, and Thomas was the rapacious ex-boss who lied his way onto the Supreme Court.

On November 3 the Thomases moved into their new house. That night "the Year of the Woman" entered the political record books.

Capping a stunning campaign against an incumbent president, Governor Bill Clinton of Arkansas defeated President Bush, returning a Democrat to the White House for the first time in twelve years. Voters' dissatisfaction with Bush's handling of the economic recession played a major role. But women voters also claimed part of the credit for Clinton's victory. Four women were elected to the U.S. Senate. Almost half the women running for the House also won, a record showing in the other half of the Capitol.

Thomas, meanwhile, felt conflicted about how to rehabilitate his

public image. Friends urged him to speak out more to counteract the
negative attacks of his critics. But Thomas resented being told what to
do. "People told him he had to go out and make his case," said Ricky
Silberman. Thomas's response was blunt: "I'm not making any case.
I'm here, let them make their case. Let them read my opinions. I say
what I have to say in my opinions, and I live my life the way I should
live my life, and they can go to hell."

On a deeper level, however, Thomas knew his friends were right.

"Thomas is very conscious of his public image," Armstrong
Williams said. "He knew he could not isolate himself and not speak
out." The question was how to try to turn things around, and under
what circumstances.

In May 1993 Thomas ventured into the public spotlight for the
first time since his confirmation. The invitation came from Mercer
University in Macon, Georgia, a conservative law school in a conservative region of the country. The parameters of the event were to become standard for all of Thomas's future speaking engagements: no
interviews with the media and no questions from reporters afterward.
Later Thomas would also stipulate at such events that questions had
to come from students.

Thomas arrived at the Law Day lunch on crutches. Two weeks earlier he had torn his Achilles tendon playing basketball against his
clerks in the gym on Court's top floor, dryly known as the "highest
court in the land." (Thomas was reportedly winning at the time of his
injury.)

The speech sounded two major themes that Thomas would return
to many times: civility in public debate and the double standard of
race.

Thomas saw opposition to his nomination as punishment for his
refusal to toe the party line on race in America, a punishment far
harsher if you were black. Whites were simply dismissed as racists for
opposing programs such as affirmative action. Blacks were guilty of
high treason. "During the 1980s I watched with shock and dismay

how friends were treated for merely disagreeing with what my friend Tom Sowell referred to as the 'new orthodoxy,'" he said. "As a black person, straying from the tenets of this orthodoxy meant that you were a traitor to your race. You were not a real black, and you were forced to pay for your ideological trespass—often through systematic character assassination, the modern-day version of the old public floggings. Instead of seeing signs on public doors saying 'no coloreds allowed,' the signs I saw were 'no non-conventional ideas allowed.'"[7]

Thomas accurately captured some of the hostility toward him from civil rights organizations, but he failed to acknowledge the gender politics that also drove opposition to his confirmation. Women's organizations, motivated by the abortion issue and Hill's charges of sexual harassment, exerted far more pressure on the Senate and public opinion than did civil rights groups.

The speech also revealed a fundamental truth about Thomas. As much as he promotes a color-blind reading of the law, he still views the world through the prism of race—especially when he is attacked.

On June 14, 1993, President Clinton nominated Ruth Bader Ginsburg, a friend of both Thomas's and Scalia's from the D.C. circuit court to replace Justice Byron White. Ginsburg was Clinton's fourth choice after three white men, including New York governor Mario Cuomo, turned the job down.[8]

In the Senate the Democrats scheduled the confirmation hearings for July, before the summer recess. Under the category of "lessons learned," the committee agreed to review and investigate any personal allegations leveled against the nominee in closed, executive session. None were made. Ginsburg's confirmation hearings lasted only four days, and she won unanimous support from the eighteen-member Judiciary Committee. She was confirmed with only three no votes from Republican senators, the first Jewish justice since Abe Fortas. Ginsburg took her place on the high court and quickly aligned herself with its more liberal wing—Justices Blackmun, Stevens, and Souter, all Republican appointees.

Ginsburg's appointment meant Thomas was no longer the Court's junior member, a distinction that carries all sorts of ancient custom at the tradition-conscious Court. Thomas shifted to the opposite end of the Supreme Court bench, reflecting his bump in seniority. In conference, the closed-door meetings where the nine justices meet to vote on cases, Thomas was no longer responsible for answering a knock at the door, a job reserved for the Court's junior member. The change also removed a little bit of the pressure that comes with being the least experienced justice on the bench.

Outside the courtroom, however, Thomas's notoriety was escalating sharply. In November his face appeared on American newsstands dressed up as Aunt Jemima, the kerchief-headed black mammy whose image was synonymous with subservience to the white "massah." The cover of *Emerge*, a magazine targeted at the black upper class, accurately captured the tone of what was inside. "Malcolm X, if he were alive today, would call Thomas a handkerchief head, a chicken-and-biscuit-eating Uncle Tom," director Spike Lee said in the piece, entitled, "Doubting Thomas." Margaret Bush Wilson, the NAACP leader who had hosted Thomas in her Saint Louis home in 1974, said Thomas wasn't "the Clarence Thomas that I knew. He's just a follower of [Justice Antonin] Scalia." A professor of African-American studies who had never met Thomas described him as "the worst kind of racist—a black man who hates himself."[9]

In the preceding Court term Thomas had cast two votes that were particularly odious to the civil rights establishment. In *Shaw* v. *Reno*, Thomas sided with the majority and allowed white plaintiffs to challenge North Carolina's congressional map with a claim of racial gerrymandering. The Court suggested new, tougher standards for how state legislatures could use race to draw political boundaries. The case grew out of the Justice Department's efforts to force North Carolina to create at least two majority-black voting districts. The case sharply divided the nine justices. Drawing districts with constituents "who may have little in common with one another but the color of their skin bears an uncomfortable resemblance to political apartheid," the Court majority said. The decision was a major setback for civil rights advocates, who believed such districts were the only way to elect black representatives, particularly in regions historically hostile to them. As the Court's only racial minority, Thomas was seen as the fifth and deciding vote in the 5 to 4 decision.

The other case that drew ire from civil rights leaders involved a death-penalty statute from Texas. Thomas joined the Court in upholding the constitutionality of the law's sentencing guidelines. His vote was a clear break from Marshall, perhaps the Court's most outspoken opponent of the death penalty. Marshall believed the death penalty was never applied fairly against American blacks, and he voted against its provisions—and for death-penalty appeals—whenever possible.

Here again, Thomas's method of constitutional interpretation prevented him from reaching the same conclusion Marshall reached. The death penalty isn't mentioned in the Constitution, but its existence is implied in the Constitution's provision that no American can be denied "life" without due process of the law. Thomas interprets such references to mean that the framers sanctioned the death penalty and that it can't be categorically ruled unconstitutional. For Thomas the death penalty was potentially discriminatory against black defendants if there were no clear guidelines for juries to follow in imposing a sentence of death. In his concurring opinion, Thomas said that the Texas statute was constitutional because it allowed defendants to present mitigating evidence in their favor.

In June 1994 the Court handed down another major decision on the 1965 Voting Rights Act. The case centered on a concept called "vote dilution," any practice of electing representatives or qualifying voters that weakened or diluted the ability to vote. The case, *Holder* v. *Hall*, came out of Thomas's native Georgia, brought by a group of black plaintiffs living in a rural county. They charged that the county's practice of electing a county commission with just one member deprived black voters the opportunity to elect commissioners of their own choosing. They argued that a five-member commission structure would allow the creation of majority-black districts and the election of black commissioners. By a 5 to 4 vote, with Thomas in the majority, the Court upheld the county's single-member commission.

Thomas's lengthy concurring opinion underscored the extent of his influence on the Court's thinking in *Shaw* v. *Reno* the year before. He said the creation of majority-minority districts segregated Americans into "political homelands that amounts, in truth, to nothing short of a system of 'political apartheid,'" quoting the words that first appeared in the majority's opinion in *Shaw*. "Blacks are drawn into 'black districts' and given 'black representatives'; Hispanics are drawn into Hispanic districts and given 'Hispanic representatives'; and so on," Thomas said. "The assumptions upon which our vote dilution decisions have been based should be repugnant to any nation that strives for the ideal of a color blind Constitution."

Thomas argued that the underlying principle of racial gerrymandering was that Americans of the same race must all think alike and could only be represented by a member of their own race. "Our drive to segregate political districts by race can only serve to deepen racial divisions by destroying any need for voters or candidates to build bridges between racial groups or to form voting coalitions," he wrote. "'Black preferred' candidates are assured election in 'safe black districts'; white preferred candidates are assured election in 'safe white districts.' Neither group needs to draw on support from the other's constituency to win on election day."[10]

The opinion captured beliefs that had been years in the making. Thomas saw the plan to create black voting districts as government-sponsored segregation, the very system he had lived under as a child. He also flatly rejected the assumption that black voters all thought alike and could only be adequately represented by another member of their own race. The opinion was also striking for his use of the term "color blind Constitution," an idea he had espoused repeatedly during the 1980s. Now, as a Supreme Court justice, Thomas was writing the idea into the law, building on Justice Harlan's dissent in *Plessy* a century ago.

Thomas's opinion prompted an outcry from civil rights leaders. The ruling struck at the heart of the concept of bloc voting, which the NAACP had employed repeatedly to pressure white southern politicians. In the decades leading up to the civil rights movement, black voters had only won concessions from white politicians when they voted together in blocs. Indeed, bloc voting prompted Savannah city officials to hire the first black police officers in 1947, the year before Thomas was born.

Conversely, many election laws in the South were geared toward limiting black voting strength. Many Georgia cities and counties abandoned multimember political districts in favor of single commissions, or at-large voting, precisely to eliminate the chance of black voters electing black representatives.

Thomas's opinion drew an unusual reproach from four of his colleagues. Justice Stevens, one of the four dissenters, penned a separate opinion that took issue solely with Thomas's concurrence. Stevens accused Thomas of a "radical reinterpretation" of the Voting Rights Act, lecturing him on the act's legislative and judicial history. He also accused Thomas of seeking to substitute the law with his personal political philosophy. "Justice Thomas would no longer interpret the Act to forbid practices that dilute minority voting strength," Stevens wrote. "To the extent that the opinion suggests that federal judges have an obligation to subscribe to the proposed narrow reading of statutory language, it is appropriate to supplement Justice Thomas's writing with a few words of history."[11]

Ever so perceptibly, however, the Supreme Court was fundamentally rewriting the use of race in the law. And the man leading the revolution, the "radical" in the Court's midst, was Clarence Thomas.

CHAPTER 16

REHABILITATION

By 1994, Thomas's public and private images were beginning to diverge sharply. Court watchers were starting to note Thomas's silence during oral arguments. He rarely asked questions, a trait that became even more obvious next to the inquisitive Justice Ginsburg. His silence fueled speculation that he wasn't interested in the Court's business and created the image of Thomas as a dour, brooding man. The unspoken subtext was that Thomas wasn't as smart as his more loquacious colleagues. The criticism from civil rights organizations, combined with the "youngest, cruelest justice" moniker from *The New York Times*, was also creating the public impression that Thomas was insensitive if not hostile toward minorities.

In private, Thomas was earning just the opposite reputation. He made a point of introducing himself to every employee at the Court, from cafeteria cooks to the nighttime janitors. He played hoops with the marshals and security guards. He stopped to chat with people in the hallways. Clerks say Thomas had an uncanny ability to recall details of an employee's personal life. He knew their children's names and where they went to school. He seemed to see people who would otherwise go unnoticed. Stephen Smith, a former clerk, recalls an instance when Thomas, on a tour of the maritime courts in 1993 or 1994, was talking to a group of judges. "There was this old woman standing there in one of those blue janitor's uniforms and a bucket, a black woman," Smith recalled. "And she was looking at him, wouldn't dare go up and talk to this important guy. He left the judges there, excused himself, and went over to talk to her. He put out his hand to shake her hand, and she threw her arms around him and gave him a big bear hug."

Among his eight colleagues, Thomas was similarly outgoing and gregarious. Justice Ginsburg said Thomas sometimes dropped by her chambers with a bag of Vidalia onions from Georgia, knowing that her husband was a devoted chef. "A most congenial colleague," said Ginsburg of Thomas.

Like Marshall, who mesmerized colleagues with stories of his lawyering days in the South, Thomas became the Court's raconteur, regaling his colleagues with stories about his grandfather and growing up in Savannah. "Clarence's life story . . . is inspiring," said Ginsburg. "He has overcome more obstacles by far than the rest of us."

Thomas took an especially keen interest in his clerks and often developed an almost paternal relationship with them. He invited them home for dinner—something he hasn't done with his colleagues—and held annual reunions in his backyard. Thomas adopted an open-door policy in chambers and encouraged his clerks to come to him with personal or professional problems. "There is still a little of the minister left in him," said Helgi Walker, who clerked for Thomas in the mid-1990s.

Outwardly, Walker had little in common with Thomas. She was born to a life of privilege in Atlanta and breezed through the University of Virginia Law School, graduating at the top of her class. The success, however, belied a life full of personal struggle, stemming from a particularly painful divorce between her mother and father. "I wasn't the victim of discrimination," she said. "But bad things can happen to all children, and he and I really bonded over that."

Thomas showed his affection for his clerks by taking an interest in all aspects of their lives, however mundane. When he noticed the treads on Walker's car were thin, he showed her how to measure them for wear and tear. "The next Monday," Walker recalled, "he came in and said, 'I saw some great tires at Price Club, they're a good deal. You should really get them.' And I'm sitting there thinking, here's a Supreme Court justice who's worried about whether my tires are safe."

Many of Thomas's clerks have similar stories to tell. One year he lent his car to a clerk when the clerk's car broke down. When the wife of a former clerk developed lupus, Thomas called her every other day at the hospital in Baltimore and followed up with a three-hour visit. When the clerk took a teaching job at nearby George Mason University, Thomas offered him a spare bedroom where he could sleep on the days he was teaching.

Walker said a common thread runs through all Thomas's interactions with his clerks: "He loves people who have refused to let whatever their circumstances are dictate their lives," she said.

In the fall of 1994 Thomas began a public relations offensive among black leaders and opinion makers. With Armstrong Williams's help, he arranged to meet some thirty black journalists, businesspeople, and political figures at the Court. Despite what Thomas said publicly, he cared deeply about how he was perceived in the black community. The tenor of the meeting was cordial, but Thomas took some pointed questions from reporters. The *Washington Afro-American* asked Thomas if he was an "Uncle Tom" or a "sellout to White people."

"I am not an Uncle Tom," Thomas said. "I do not pay attention to that nonsense ... I have not forgotten where I come from ... I feel a special responsibility to help our people, and I am doing the best I can."[1]

Thomas also reached out to prominent black athletes, particularly basketball players. He counseled Charles Barkley about the latter's highly publicized outbursts at fans and referees, saying, "Son, if you keep giving them the hammer to hit you with, sooner or later they're going to hit you too hard," Thomas told Barkley, according to Barkley.[2]

In 1993, conservative author David Brock published *The Real Anita Hill*, which cast Hill as an unstable woman with a political grudge against Thomas. Although Brock later recanted some of the book's most salacious details in *Blinded by the Right*, published nine years later, Thomas's supporters believed Brock vindicated Thomas when his first book was published.

But Thomas continued to fight a mostly losing public relations battle. His name and reputation were subjected to even more public damage when yet another book about his confirmation was published in November 1994. *Strange Justice*, written by two veteran Washington correspondents, portrayed him as a calculating opportunist who waged a decadelong campaign to reach the high court. The book suggested that he lied about what happened between him and Anita Hill, quoting other women who charged Thomas with making lewd comments toward them. The book also portrayed Thomas as obsessed with pornography, interviewing video-store clerks who alleged that Thomas rented X-rated films. The book received widespread play in the media and sold briskly.

At the same time Danforth published *Resurrection*, his own account of the confirmation battle. The book, which chronicled the confirmation saga's emotional toll on Thomas, sold poorly.

The new books prompted a spate of new polling that showed Thomas's standing among Americans had dropped. Respondents to a *Newsweek* poll said they believed Hill by a margin of two to one. The poll respondents divided equally over whether Thomas should have been confirmed, a steep drop-off of support. Try as he might, Thomas could not seem to put the confirmation behind him.

The 1994–1995 Supreme Court term became the most prolific and productive of Thomas's tenure. Now more comfortable on the bench, he was beginning to hone a system for deciding cases and writing opinions. He spent long hours researching the history of laws at issue, reviewing all of the Court's precedents. Thomas's mind worked linearly and logically. He had to understand how a thing had come about before he could determine where it should go.

"His first question always is, How did we get here?" said Helgi Walker. "We would go back through that entire history, and if there was a point at which the justice thought a wrong turn had been made, he would willingly say that."

Judicial decision making, particularly at the Supreme Court, is an intricate, complex undertaking. Conservative jurists tend to regard the Constitution as a static document that means exactly what it says. Liberal jurists tend to regard it as a statement of elastic principles that were meant to grow with the nation. Supreme Court precedents also figure into the decision-making process. What has the Court said before, and what does that say about the case under review? The justices can also look to legislative history, or intent, when considering the validity of a particular law. Then, of course, there are the individual life experiences that each justice brings to the table. The predilections act like lenses through which the justices examine the other available guideposts.

As a conservative, Thomas was drawn to constitutional textualism, or reading the law for its plainest, most obvious meaning. The approach fit neatly with Thomas's background in the segregated South. Like Myers Anderson, Thomas saw the world in absolutes. Thomas grew up in a household where right was right and wrong was wrong.

Thomas also grew up in an environment where the rules were different, depending on your race. As a black boy in the South, your safety depended on your ability to learn those rules and never break them. Don't fool with white women. Don't go to white neighborhoods. Don't talk back to white folk. Above all, don't go anywhere where you don't know the rules. Part of the reason Anderson never took his family outside Chatham, Bryan, or Liberty counties was because he knew the rules in those places. Anywhere else was dangerous.

As a justice, Thomas pays close attention to rules. He likes bright line rules, the clearer the better, and the same for everybody. He doesn't like creating exceptions or special rules to fit a particular need. Thomas also believes that legislatures, not unelected judges, make laws. "It is not for us to amend the Constitution, to rewrite the Constitution, to make over the Constitution or be philosopher kings," Thomas told one audience. "We are judges who interpret what the political branches enact, who interpret what the framers wrote."

Thomas said there is only one way to determine the intent of the Constitution's framers: the plain reading of the words. "It is either there or was intended to be there, or it isn't," he said. "It's as simple as that."[3]

Thomas says his approach limits the possibility that he can overlay his Supreme Court opinions with his own personal biases. He never looks at a dispute and asks himself how he can fix it. He focuses instead on the principles at stake, and how a particular case will govern other sets of circumstances.

The integrity of his methodology becomes murkier when the cases involve issues close to Thomas's heart, such as race and the law, or religion in school. Thomas writes deeply and passionately about both issues, expressing ideas torn straight from his youth.

Thomas's four law clerks are also a central part of his system. He requires that his clerks be in at least the top 10 percent of their law school graduating class. And he wants good, clean writers. Thomas thinks many of the Court's opinions are too complicated, telling his clerks he wants decisions that even a gas-station attendant could read and comprehend.

Thomas also likes young people with personality, who aren't afraid to express themselves and mix it up with him. "He wants free thinkers," said John Yoo, who clerked for Thomas in 1994 and 1995.

"He wants people to speak up with views that may be wacky or haven't been argued for one hundred years, but he wants to hear the whole range of views."

Over the years Thomas has hired men and women from a mix of different backgrounds, including one openly gay clerk.

There are limits to Thomas's free thinking, however. Thomas only hires clerks who share his conservative philosophy.

At the start of every term, Thomas requires his four clerks to sit through a screening of his favorite film: Ayn Rand's *The Fountainhead*, the same movie that Thomas identified so strongly with at the EEOC. "The justice makes you watch it, and then you have to groan, and say how bad it is," said Walker. "Then he laughs. But there is a message in all of that, and the message is: Do not let other people change what you think is right."

Thomas gives clerks only two guidelines in preparing opinions. They are never to bargain points of law to win the support of other justices, and they may never take personal shots at another justice. Ideas are open to criticism, but never the authors. "He would never permit a draft to come to his desk that had disparaging remarks about another justice," said Greg Coleman, who clerked for Thomas in the mid-1990s.

Thomas developed a routine for preparing for oral arguments. He read all of the briefs before the case was to be heard, usually over the summer before the start of the term. The morning of the argument, he met early with his four clerks in his chamber. They sat around his coffee table, munching from a bowl of candy that Thomas kept stocked for the occasion. Then he listened to his clerks debate each case. His guideline was simple: He didn't want to hear anything in oral argument that he hadn't already heard that morning. He encouraged a free-form debate, often withholding his own views until the very end.

The Supreme Court receives about eight thousand appeals each year and accepts about eighty for review. The cases fall into many categories, but the justices pay particular attention to those where lower courts have disagreed about a law's constitutionality or there is a question about what Congress intended the law to do. They also take cases where there is a clear question of the law's compatibility with the U.S. Constitution. Once the term gets under way, the Court sits for three days at a time about every other week to listen to oral arguments in each case. Except in extreme cases, the arguments last one hour, equally divided among the two parties.

Once or twice a week the nine justices gather to cast their votes on the cases argued for the week. They sit at a long rectangular table. Rehnquist sits at one end, Stevens at the other. Scalia, O'Connor, and Kennedy are on one side of the table; Thomas, Ginsburg, Souter, and Breyer (who replaced Blackmun in 1994) are on the other. The chief justice introduces each case and casts the first vote. Voting then proceeds by seniority, with any justice free to argue points of law.

Justice Scalia says Thomas, though reticent on the bench, is an animated and passionate debater in conference. Thomas takes particular interest in cases involving employment law, his specialty at EEOC, and issues close to his heart, such as affirmative action. "He is at all times well-prepared for our conferences," said Justice Ginsburg, "and routinely presents well thought-out statements of his position."

After voting on cases, the chief justice assigns the writing of the opinions if he is in the majority. If not, the most senior judge in the majority assigns the case. Any justice is also free to write a separate concurring opinion to flesh out his or her ideas. Dissents are written in the same fashion, either individually or in a group. Both dissents and concurrences indicate that the author feels particularly strongly about that case and the legal issues involved.

Thomas frequently writes concurrences or dissents that stake out legal ground far beyond what any of his colleagues will embrace. Indeed, he is the most willing of all his colleagues to overrule precedent, what is known in legal jargon as *stare decisis*, or "let the decision stand," says Justice Scalia. "He does not believe in *stare decisis*, period," Scalia says. "If a constitutional line of authority is wrong, he would say let's get it right. I wouldn't do that."

A 1995 case proved the point. In April the court struck down a 1990 law banning the possession of guns within one thousand feet of a school. Although the case was technically about firearms, the real issue was congressional power, a debate that had been raging for two decades. Congress enacted the Gun Free School Zones Act of 1990 under the Constitution's commerce clause, which gives it the power to "regulate Commerce with foreign Nations, and among the several States, and with the Indian Tribes." Conservatives, including Rehnquist, had been arguing for years that Congress was using the commerce clause to regulate activities properly left to the states. Conservatives on the Supreme Court were looking for a case to rein Congress in, and they found it in the Gun Free School Zones Act. In his majority opinion, Rehnquist argued that regulating guns near school yards had nothing to do with commerce.[4]

In his concurrence, Thomas went a step farther, suggesting that a nearly sixty-year-old test used to evaluate laws enacted under the commerce clause ought to be thrown out. It was a brash critique of Court precedent and established Thomas as perhaps the Court's boldest conservative.

The opinion underscored Thomas's libertarian antipathy toward governmental power, a suspicion he developed in Georgia watching southern whites use the law to benefit themselves. At the EEOC Thomas clashed repeatedly with Congress, which he thought abused its power by coercing government agencies to adopt policies that should have been adopted by legislation.

As a justice, Thomas believes the framers of the Constitution specifically listed the powers of Congress because they did not want it to legislate every aspect of American life. It's the essence of the federalism principle articulated in the 1995 decision on guns at school, a concept that the Court has since applied to other areas.

Another case underscored just how sensitive Thomas remained to anything touching on racial discrimination and the legacy of segregation.

Lawyers for the Ku Klux Klan came before the Supreme Court in April 1995, arguing that the Klan had a First Amendment right to display a cross on the grounds of the Ohio state capitol over Christmas. The normally reticent Thomas stunned those in the courtroom with a barrage of questions directed at the Klan's lawyer. "You say that this is a religious symbol?" Thomas boomed skeptically. "What is the religion of the Klan?"

What if the cross were burning? Thomas wanted to know. What kind of symbol would it be then?[5]

Despite his skepticism, Thomas voted with the majority to permit the Klan to display its cross. He noted in his concurring opinion that a plain reading of the First Amendment clearly protected such forms of "religious" expression. But he made no effort to conceal his personal feelings about the hate group. "The cross is a symbol of white supremacy and a tool for the intimidation and harassment of racial minorities, Catholics, Jews, Communists, and any other groups hated by the Klan," he wrote.[6]

Thomas concluded the 1994–1995 term with three major opinions that clearly reflected his own personal views—two of the cases involved race, and one involved religious activities at public schools.

In a major decision on affirmative action, Thomas joined the Court in articulating a new, tougher standard for governments to justify policies based on racial classifications. The case challenged a contract incentive for road builders that used minority subcontractors in Colorado. For the first time the Court ruled that such federal policies were almost always unconstitutional unless they were narrowly tailored to redress proven discrimination. "The Constitution protects persons, not groups," O'Connor wrote for the Court majority. "It follows from that principle that all governmental action based on race . . . should be subjected to detailed judicial inquiry to ensure that the personal right to equal protection of the laws has not been infringed."

The case raised the demons of racial inferiority for Thomas, who saw in the set-aside program an implicit assumption about the abilities of minority contractors. "Racial paternalism and its unintended consequences can be as poisonous and pernicious as any other form of discrimination," he said in his concurrence. "So-called 'benign' discrimination teaches many that because of chronic and apparently immutable handicaps, minorities cannot compete with them without their patronizing indulgence.... These programs stamp minorities with a badge of inferiority and may cause them to develop dependencies or to adopt an attitude that they are 'entitled' to preferences."[7]

Despite the legal verbiage, Thomas spoke directly to the story of his own life. Twenty years earlier he believed he had been tagged with a "badge of inferiority" when he was admitted to Yale under the school's first official affirmative action policy.

"I think that's the thing he hates most about affirmative action, that it takes away from the accomplishments of those who can make it in any league," Scalia observes. "You go out and people say, Yeah, sure you did this, but who knows? You might just be the 'show black' or the 'affirmative action black.' "

Thomas amplified his views in another case decided the same day. A sharply divided Court swept aside a lower-court ruling that had forced Kansas City to spend more money on improving conditions for the city's mostly minority school system. The lower court ordered the expenditures to make the inner-city schools more attractive to suburban, white students outside the school district, and integrate the city schools.

Thomas launched a blistering attack on the foundation of the lower court's ruling, which argued that Kansas City's predominantly black and white school districts resulted from the lingering effects of segregation. " 'Racial isolation' itself is not a harm; only state-enforced segregation is," Thomas said. "After all, if separation itself is a harm, and if integration therefore is the only way that blacks can receive a proper education, then there must be something inferior about blacks. Under this theory, segregation injures blacks because blacks, when left on their own, cannot achieve. To my way of thinking, that conclusion is the result of a jurisprudence based upon a theory of black inferiority."[8] Once again Thomas mined his personal resentment over society's assumptions about black Americans to fashion Supreme Court law. Thomas labored too long under the presumption of "black inferiority" to tolerate government programs that he believed perpetuated the presumption.

Thomas rounded out the 1995 term with another major ruling that served to relax, ever so slightly, the prohibitions against public financing of religious activities. The Supreme Court, in another 5 to 4 decision, reversed a lower-court ruling that had served to withhold student-activity monies at the University of Virginia for a Christian student group. The Court held that the payments would not violate the First Amendment's prohibition against state-sponsored religion if the funds were equally available to all groups. The First Amendment's so-called establishment clause "does not compel the exclusion of religious groups from government benefits programs that are generally available to a broad class of participants," Thomas wrote in his concurrence.[9] Thomas made no references to his own parochial-school background. But his views were closely tied to the dominant role that the church played in his life. Outside the Court, Thomas's concurrence in the Virginia case became the legal foundation for the school-voucher movement, giving religious conservatives new hope that taxpayer-supported vouchers might one day be allowed in parochial school.

In the late spring of 1996 Thomas returned to Worcester to attend his twenty-fifth-year reunion at Holy Cross. Among his former Jesuit teachers, including Father John Brooks, Thomas disclosed publicly a profound change in his life: He had rejoined the Catholic church after a twenty-eight-year estrangement. Thomas's return to the church was years in the making and involved years of soul searching. His separation from the church in 1968 after King's assassination had been more than just an abandonment of Catholic teachings. It was a rebellion against his grandfather and the rigid authority that defined so much of his early life. Thomas believed that the church had broken its promise to him when he came to believe that it was no less bigoted than any other institution in American life. Forgiveness came hard for Thomas. Trust once broken was never easily repaired.

Although Thomas had begun attending morning mass every day at a Washington parish following his grandparents' deaths in 1983, he had not rejoined the church. Then he met his wife and began attending an Episcopalian church in Virginia with her. Then came the confirmation ordeal, which brought him face-to-face once again with the power of faith. Over those days in October he was at times crushed under the weight of shame and guilt. The prayers of people like John Danforth and Charles James gave him the strength to go on.

After he joined the Supreme Court Thomas discovered a small Catholic church just two blocks away. Founded in 1868 to serve Washington's growing community of German Catholics, Saint Joseph's on Capitol Hill in the early 1990s was a quiet sanctuary for Catholics working in the many government buildings around it. Senator Kennedy worshipped there, along with Justice Scalia. Thomas developed the habit of attending the church's eight o'clock mass virtually every day. At home on Sundays he continued to attend an Episcopal church with his wife. His epiphany came one Easter Sunday. Driving home from the service, Virginia Thomas asked him how he'd liked it. "It was wonderful," he told her. And then he blurted out, "But I'm Catholic."[10]

The year 1996 also marked Thomas's fifth anniversary on the Court. Major news organizations weighed in on Thomas's tenure and impact in the months leading up to the event. Legal scholars noted his role in the Court's redefinition of race and the law. Thomas was also viewed as a key player in the Court's emerging federalism. The reviews were mixed, depending on the political persuasion of the reviewer. Some analysts continued to suggest that Thomas was nothing more than a stepchild of Scalia, noting the frequency with which the Court's two most conservative members voted together. And some angrily noted that Thomas was hardly acting like the moderate who promised during his confirmation hearings to tread cautiously over established Supreme Court precedents.

Yet there were also more independent assessments. "His work goes a long way to refuting the notion that this is someone who does not have the breadth to be a Supreme Court justice," Richard Lazarus, a law professor at Washington University, told the *St. Louis Post-Dispatch* in Thomas's former home state of Missouri in 1995. "One can disagree with him and think he is misguided, but you cannot read these opinions and think this is someone who does not have the command of legal argument."

Ideologically, he was most closely allied with the chief justice and Scalia. Together the three jurists formed the Court's more conservative bloc, although Thomas and Scalia were the more conservative of the three. Justices Kennedy and O'Connor tended more toward the Court's middle but sided with their conservative colleagues on key cases. The Court's more liberal wing included Justices Stevens, Souter, Ginsburg, and Stephen Breyer.

Consistency was also emerging as a hallmark of Thomas's jurisprudence. Thomas almost never departed from his conservative judicial principles; the consistency was calculated, however. As a student of history, Thomas determined that his Supreme Court tenure would ultimately be judged not case by case but as an entire body of law. "Thomas's view is that any respect he'll get is after he's dead," said John Marini, who advised Thomas at the EEOC.

Thomas's judicial decision making drew accolades from conservative legal scholars, who praised him as the boldest member of the

Court's conservative wing. Thomas was also wowing the conservative rank and file.

In 1996 Phyllis Schlafly, president of the conservative Eagle Forum, was nearly breathless in her praise of Thomas. "George Bush was right when he said Clarence Thomas was the most qualified man to appoint," she said. "He's better than Rehnquist. He's better than Scalia. He's just wonderful!"

As Thomas's star rose among conservatives, however, his every move seemed to generate controversy in the black community. In May 1996 a predominantly black school in Prince Georges County, a Maryland suburb of Washington, rescinded an invitation to Thomas to speak at an awards ceremony for black youth after a school board member complained. "These are the youngsters who are having problems in our schools," explained Kenneth Johnson, the board member who objected to Thomas's visit. "They are African-American males. Justice Thomas simply is not the role model that I would like to see for them to have."[11] A majority of other board members, however, disagreed, and Thomas was reinvited.

A group of some eighty parents, joined by the district's black member of Congress, held what amounted to a counterdemonstration while Thomas spoke. They carried signs reading, UNCLE THOMAS IS A TRAITOR and NO UNCLE TOM IN OUR COUNTRY.

Thomas's stock also continued to plummet among black intellectuals and writers. For the second time in three years *Emerge* magazine weighed in with a blistering critique. This time the magazine's cover featured Thomas as a black lawn jockey, one of the most hateful symbols of southern bigotry in the American stockpile of racial symbols. The accompanying text was even more biting than the magazine's first article on him in 1993.

The article castigated Thomas for his votes on the Colorado affirmative action program and the voting-rights cases that had come before the Court. "I do not consider him to be a positive role model for Blacks," the legendary Rosa Parks told the magazine. "He had all the advantages of affirmative action and went against it. I feel he should have taken up the responsibility of [Thurgood] Marshall, but since he did not, I can't think of anything we can do to change his mind."

Joseph Lowery, still smarting from the SCLC's decision not to oppose Thomas in 1991, urged the magazine's readers to pray for Thomas. "He is becoming to the Black community what Benedict Arnold was to the nation he deserted; and what Judas Iscariot was to Jesus: a traitor, and what Brutus was to Caesar: an assassin!"[12]

Despite the vitriol, the critiques captured a widespread sentiment among blacks: Thomas doesn't empathize with the hardships that many continue to face in a majority-white society. Thomas's faith in color-blind laws—so obvious to him—simply doesn't ring true for those who feel the playing field still isn't level.

Earlier in the year *Emerge* had held a roundtable discussion on affirmative action with some of the nation's leading scholars on African-American history. Manning Marable, a Columbia University history professor and director for the Institute for Research in African-American Studies, said Thomas's conservative philosophy made him no longer black. "Even though he is black in terms of his racial identity, the Clarence Thomas in terms of his political program, in terms of his repudiation of civil rights, is arguably, if you use the racial discourse, the Whitest man in America."[13]

Outwardly, Thomas professed indifference to the increasingly hyperbolic attacks against his racial identity and conservative jurisprudence. Some supporters printed up bumper stickers after the Maryland school protest urging Washingtonians to boycott Prince Georges County. Thomas gamely hung one in his Supreme Court chambers.

Inwardly, however, the attacks on his blackness rankled Thomas. Rather than ignore his critics, he resolved that it was time to confront them head-on. Yet Thomas was a cautious man, and he wanted time to think about what he wanted to say and when he wanted to say it. As he had told the students at the school in Prince Georges County: If you don't have anything nice to say, don't say it.

The opportunity Thomas was looking for came in the summer of 1998. The National Bar Association, the organization of black lawyers that had debated for more than seven hours on his nomination in 1991, invited him to speak at their annual meeting. The event was scheduled for the last week in July in Memphis, the thirtieth anniversary of Martin Luther King's assassination. For Thomas, a convention of black lawyers and judges was just the platform he wanted to answer the mounting criticism of his Supreme Court decisions.

Predictably, Thomas's invitation generated heated debate within the NBA, with a vocal contingent of prominent members urging that the offer be withdrawn. A. Leon Higginbotham, a retired federal judge who had served on the Yale Corporation while Thomas was in law school, was outraged that the NBA was giving Thomas such a prominent platform to express his views, and at a time and place so closely tied to the remembrance of King. The retired judge convinced a majority of the judiciary council's executive board to rescind the invitation, but its chairwoman, Louisiana Supreme Court justice Bernette J. Johnson, refused to comply with the ultimatum.

The bid to uninvite him was just the sort of pressure that guaranteed Thomas wouldn't back out. He would not be intimidated. He respected people who disagreed with the merits of his Supreme Court rulings, but he could no longer abide the constant assaults on his race and political views.

Higginbotham and others opposed to Thomas arranged to hold a panel discussion on civil rights at the convention, deliberately timed to coincide with Thomas's luncheon speech. They hoped to embarrass the Supreme Court justice by enticing some luncheon guests to walk out of the ballroom as he spoke.

Thomas spent hours drafting and revising the speech. Among other sources, he turned for guidance to his literary mentor Ralph Ellison, whose *Invisible Man* had been so meaningful to him in college.

"The World and the Jug," an essay, particularly inspired him. The essay was a manifesto for intellectual freedom, written in response to Irving Howe, a Jewish intellectual from New York who criticized Ellison for failing to fulfill the militant literary legacy of Richard Wright. In his response Ellison likened Howe's literary prescription for black authors to white supremacists in Mississippi; both were seeking to confine black Americans into prisons of their own choosing. The experiences of black Americans were hardly monolithic, Ellison wrote. Ellison asserted that he was as much a "Negro" writer as an *"American writer."* "Being a Negro American involves a willed . . . affirmation of self against all outside pressures," he wrote.

"Everybody wants to tell us what a Negro is, yet few wish, even in a joke, to be one," Ellison wrote. "But if you would tell me who I am, at least take the trouble to discover what I have been."[14]

In Memphis Thomas planned to tell the nation who he was, and what he had been.

Thomas was pumped for the speech as he flew to Memphis. Shortly before he was to speak, one of the introductory speakers invited audience members to visit the rest rooms, a signal to those protesting to make their move out. A handful of luncheon guests took the cue.

Thomas looked grim and resolute as he began speaking. He included no jokes or crowd pleasers to soften up his audience. He was there to stand up for himself.[15]

Stripped to its essence, the speech was a passionate plea for acceptance. Thomas flatly rejected the notion that blacks were confined to the same opinions because they shared the same skin color. To suggest otherwise, Thomas said, was as dehumanizing as the denial of rights and freedoms under segregation. Thomas strongly condemned the double standard that America applied to black conservatives such as himself. Whites who articulated conservative views were treated with respect, Thomas said, their views subject to thoughtful analysis and discussion. Blacks who espoused similar views were reviled. For the first time Thomas angrily refuted the suggestion that a "pied piper on the Court"—a thinly veiled reference to Scalia—led him and the other justices or wrote their opinions for them. Since thinking outside society's accepted norms for black Americans was "presumptively beyond my abilities," Thomas said, "obviously someone must be putting these strange ideas into my head and my opinions.

"Though being underestimated has its advantages, the stench of racial inferiority still confounds my olfactory nerves."

Thomas scored white Americans for assuming that blacks were a monolith of thought and political expression, comparing the stereotype to that of the dumb, lazy black man created under Jim Crow. "It no longer matters whether one is from New York City or rural Georgia," he said. "It doesn't matter if we come from an educated family or a barely literate one. It does not matter if you are Roman Catholic or Southern Baptist. All of these differences are canceled by race, and a revised set of acceptable stereotypes have been put in place. Long gone is the time when we opposed the notion that we all looked alike and talked alike. Somehow we have come to exalt the new black stereotype above all and demand conformity to that norm."

Reaching the climax of his remarks, Thomas defended his opposition to affirmative action, perhaps the salient issue that divided him from other black Americans. "Any effort, policy, or program that has as a prerequisite the acceptance of the notion that blacks are inferior is a non-starter with me," he said. "I do not believe that kneeling is a position of strength. Nor do I believe that begging is an effective tactic. I am confident that the individual approach, not the group approach, is the better, more acceptable, more supportable, and less dangerous one."

Abruptly, Thomas then softened his stance and revealed what he had long publicly denied: He was deeply pained by suggestions that his actions hurt blacks. "More deeply than you can imagine," he said softly. "All the sacrifice, all the long hours of preparation, were to help, not hurt."

Thomas said he came to Memphis "to assert my right to think for myself, to refuse to have my ideas assigned to me as though I was an intellectual slave because I'm black. I come to state that I'm a man, free to think for myself and do as I please.

"But even more than that, I have come to say that isn't it time to move on? Isn't it time to realize that being angry with me solves no problems? Isn't it time to acknowledge that the problem of race has defied simple solutions and that not one of us, not a single one of us, can lay claim to the solution? Isn't it time that we respect ourselves and each other as we have demanded respect from others? Isn't it time to ignore those whose sole occupation is sowing seeds of discord and animus? That is self-hatred."

Thomas was finished. He stepped back from the podium while delegates rose and gave him a tepid round of applause. Apart from polite hand clapping when he was introduced, there were no interruptions of applause during his remarks.

Back in Washington, meanwhile, the most salacious scandal since Anita Hill accused Thomas of sexual harassment was unfolding at the White House. Kenneth Starr, an independent counsel, was investigating allegations that President Clinton lied under oath about a sexual relationship with a White House intern named Monica Lewinsky. Starr delivered his findings to Congress on September 11, 1998. The 453-page report, laced with graphic details of Clinton's sex life in the Oval Office, chronicled a tawdry eighteen-month affair between the president and a woman young enough to be his daughter. Starr's allegations of lying under oath and obstructing the inquiry became the basis for the nation's first impeachment proceedings since 1868.

The scandal laid bare the deep partisan animus coursing through Washington and exposed the city's hypocrisy on matters of sex and power. Conservatives who protested so loudly about the smear campaign against Thomas were leading the pack to impeach the president for his private indiscretions. Feminists and leaders of prominent women's organizations who had called for Thomas's head during his confirmation were all but silent over the president's potentially exploitive conduct toward a young woman. The nation's capital was consumed in political warfare, obliterating the lines between public and private conduct.

Thomas seemed somehow liberated after his speech in Memphis, and he dramatically increased the frequency of his public speaking engagements, sometimes scheduling two or three a month. The pace far exceeded the speaking schedule of any of his colleagues. Reticent on the bench, Thomas was fast becoming the most loquacious justice off it.

The appearances were the first sure indications that Thomas was emerging from his shell. He picked small universities or safe, reliably conservative audiences. He continued to refuse media interviews or questions from reporters, but he encouraged C-SPAN cable television to broadcast the speeches. Thomas liked the fact that the network broadcast his remarks in full, without analysis, commentary, or packaging, allowing Thomas to speak to the nation without filters.

Often Thomas addressed issues directly relating to his confirmation. He spoke about the distorted influences of special-interest groups on the judicial nomination process. He talked about the need for civility in public dialogue. He told students to be heroes, not victims, lamenting the tendency to assign blame instead of seek solutions.

Some of the remarks were sharply partisan. For example, he vigorously defended Republicans in Congress who were attacking President Clinton's judicial nominees. The deepening crisis over judicial appointments had escalated after Clinton's reelection in 1996. The Republican-controlled Senate, accusing Clinton of appointing "judicial activists," slowed the pace of judicial confirmations to a standstill. Once again, the controversy exposed partisan hypocrisy on Capitol Hill. Conservative Republicans who were so outraged over efforts to derail Reagan and Bush appointees over political ideology were now imposing their own ideological litmus test. The same Democrats who argued that ideology was central to picking judges were now asserting ideology was off limits.

Thomas defended not the nominees but their Republican accusers. In several 1999 speeches he said criticism of the judiciary was healthy, and that judges who were intimidated by criticism had no business being judges. There was an almost petty, self-pitying tenor to the remarks. More than once Thomas noted that no one had come to his defense when he was under attack. "Judges simply do not need protection from the slings and arrows of mere words," Thomas told an audience in Montana. "We are not that fragile."[16]

Thomas also began speaking more openly and candidly about his confirmation. He told an audience in Ohio that the Senate had failed him. "I was let down by those who should have known better, who should have stood between me and the mob," Thomas said.[17] In Phoenix Thomas publicly acknowledged for the first time how deeply scarred he was after his confirmation. "I was a wounded bear," he said. He also publicly said for the first time that he had finally put the ordeal behind him. "With the grace of God and your prayers, we survived," Thomas said. "We have recovered from it."[18]

There were also less publicized appearances. Just before Christmas in 1998 Thomas spoke at a "graduation" ceremony for recovering drug addicts at a youth home in Virginia. He stood before the small gathering and urged them to put the past behind them and make something positive of their lives. It was a stunning picture, recalled Tom Jipping, a Washington conservative. "One of the nine most powerful people in the world out at a youth home speaking to a bunch of former drug addicts and gang members."

For those closest to Thomas, however, there was nothing odd about such encounters. Thomas knew all about pain and suffering, and he'd learned how to keep going. At bottom, Thomas was a survivor.

Yet his closest friends knew a part of him would never put the confirmation behind him. Privately, he continued to harbor deep resentments toward Senate Democrats, who he believed leaked Hill's charges to the media. Thomas was also still deeply hostile to the media, who he believed had failed to tell his side of the story fairly. "He's never put it out of his mind, and he never can," says Danforth. "That was a very scarring experience for him."

Thomas's increasingly frequent and publicized speeches away from the Court, however, called attention to his silence on it. Thomas's reticence was now common knowledge around the country. Thomas asked a question or two from time to time, but they were few and far between. Twenty years ago questions from any of the justices during oral argument were the exception, according to Court scholars. But the nature of the arguments changed dramatically in the 1980s and 1990s, with justices interrupting lawyers within the first minute of their presentations. (Justice Scalia says he believes he started the trend.)

In his speeches Thomas offered many reasons for his silence. He said the arguments are for the lawyers arguing their cases. He said most questions get asked anyway, so there is no need for him to ask one. He has suggested that his colleagues ask too many questions, and that some are just for show. In some of these explanations Thomas suggested that oral arguments are a mere formality. He's all but said that his mind is made up before the first word is spoken, having read and studied the briefs ahead of time.

Thomas also cited personal reasons, noting that he grew up in a household where children did not speak unless spoken to. He talked movingly of his struggles to learn Standard English as a child, and how self-conscious he felt in college and law school. "The problem was that I would correct myself midsentence," Thomas said. "I was trying to speak Standard English, I was thinking in Standard English but speaking another language. So I learned that—I just started developing the habit of listening, and it just got to be—I didn't ask questions in college or law schools, and I found that I could learn better just listening. And if I have a question, I could ask it later."[19]

Thomas's public taciturnity, however, is also rooted in his natural stubbornness. In the beginning he asked no questions because he was determined not to give his critics ammunition with which to speculate about how he might vote. He refused even when friends and law clerks urged him to ask more questions, arguing that his silence only fueled racist speculation about his abilities. Even his wife urged him to talk more. Thomas, however, was not so easily moved. "His feeling is, what does he care?" said Steve McAllister, one of Thomas's first clerks.

While Thomas ignored the innuendo about his silence on the bench, he continued to be galled by the negative comparisons to Justice Scalia, even as they became less and less frequent. Speaking to students in Louisville in the fall of 2000, Thomas fell flat with an attempted joke. Did Thomas write his own opinions? a student wanted to know. No, Thomas replied, Scalia wrote his opinions. The joke, producing an awkward silence in the auditorium, hardly concealed the anger Thomas felt at the perception that he was the hapless dupe of a white colleague. "Because I am black, it is said that Justice Scalia has to do my work for me," Thomas said earlier in his remarks. "He must somehow have a chip in my brain and controls me that way."[20]

Nine years after joining the highest court in the land, Thomas was still battling the presumption of racial inferiority that had had so much to do with defining his life.

The 2000 presidential contest between George W. Bush and Al Gore validated Thomas's emerging status as an American political icon. Asked what kind of justices he would appoint to the Supreme Court, Governor Bush said he wanted jurists in the mold of Thomas and Scalia, strict constructionists who would be faithful to the text of the

Constitution. Gore pounced on the remarks, holding up Thomas as a judicial pariah to liberal audiences, especially black ones.

But the future president's remarks confirmed what conservatives had been saying for years: Thomas was the Court's conservative conscience. Liberals seemed surprised by Thomas's sudden rehabilitation. But it had been years in the making. Thomas stuck resolutely to his conservative principles and incorporated them into everything he wrote. In 1999 Henry J. Abraham, one of the Supreme Court's foremost historians, published a revised edition of his history of the Court. "Thomas has remained steadfast in his jurisprudence and has slowly gained the respect of sundry legal scholars for his written opinions," he wrote.[21]

The same year, another Supreme Court scholar published an entire volume on Thomas's jurisprudence, an unusual distinction for a justice still so young. Author Scott Gerber concluded that Thomas's most significant impact on the Court was in the area of race and his insistence on a color-blind reading of the Constitution. Gerber also argued that Thomas was an independent voice on the Court, not Scalia's "loyal apprentice."[22]

A more partisan acclamation of Thomas's rehabilitation came two weeks after Bush succeeded Clinton as president, following the controversial election in which the Supreme Court had played so much a part. In February the American Enterprise Institute, a conservative Washington think tank, honored Thomas with a coveted award among conservatives. More than twelve hundred of Washington's A-list conservatives showed up at the Washington Hilton for the event, filling the giant ballroom and according Thomas rock-star status. Throngs waited to meet him, standing in line for more than forty-five minutes to shake the justice's hand.

Thomas's speech was a passionate call for extremism. "It is not comforting to think that the natural tendency inside us is to settle for the bottom, or even the middle of the stream," he said. "By yielding to a false form of 'civility,' we sometimes allow our critics to intimidate us. . . . This is not civility. It is cowardice."[23]

Thomas appeared to enjoy the attention, and he remained long into the night to shake the hand of every well-wisher. Only once did his demeanor change. Approached by a news reporter, Thomas stiffened perceptibly. The wide smile disappeared. He locked his arms across his chest defensively. He shifted uncomfortably before politely turning away.

Yet Thomas went home that night a winner. It had taken him more than a decade, but in the end he'd won back his dignity and validated the expectations of those who'd fought so hard to put him on the Supreme Court. With George Bush now president, there was even talk of Thomas succeeding Rehnquist as chief justice.

Few Americans glimpsed the turmoil and tragedy that swirled in Thomas's personal life, however, or understood how much of his healing had come from the gift of unexpected fatherhood and the boundless optimism of children.

CHAPTER 17

CHILDREN

B y the late 1990s, Thomas had distinguished himself in another way, one invisible to all but those who worked inside the Supreme Court. On a regular basis, sometimes as often as twice a month, groups of schoolchildren were trooping in for private audiences with the Court's only black justice. The visits began slowly and informally a few years after Thomas was confirmed. Soon word spread among school districts in the Washington area that Thomas would entertain schoolkids at the Court. Thomas preferred to talk to children who might benefit most from the exposure. The groups tended to come from predominantly black schools, and many were from poor families. Typically, he reserved one of the Court's ornate conference rooms for the sessions, blocking off a free morning or afternoon. Thomas usually opened with some words about himself, his background in Georgia, and his grandparents. He urged the children to stay in school and concentrate on their homework. Turn off the television, he invariably told them. Then he took questions. Some sessions lasted two hours or more.

Thomas had met with schoolchildren while he was at the EEOC, accepting numerous invitations to speak at local Washington schools. The interest dated back to when he tutored the kids in Sandfly while he was studying at Saint John's on the Isle of Hope.

Yet the encounters took on new meaning for Thomas after he joined the Court. Thomas liked the innocence of children. They didn't have agendas, and they didn't talk back to him. Helgi Walker remembered one particularly touching encounter from 1995.

Thomas was speaking to a group of black elementary-school students from Charlottesville, Virginia, and repeating his familiar line about turning off the television to do homework. A fifth-grader asked, "How can you tell us to do that? I bet there are things you do that you know aren't good for you." Thomas thought a moment and said, "You know what, buddy, you are right. I smoke cigars and I do that because it's fun, but I know it's not good for me. So why don't you come up here and we'll make a deal?" The boy went to the front of the room, and Thomas bent down to look in his face. "I will never smoke another cigar if you promise me that for one year, you will do your homework instead of watch television." The two shook hands on the deal.

Afterward, Thomas called Walker into his office and handed her a box of expensive cigars. "Take it away," he said. "Just take them and throw them away."

The following year a group from the same Charlottesville school returned to the Court. Thomas immediately spotted the boy who'd shaken his hand the year before, Walker said. "Buddy, how much television have you been watching?" Thomas asked.

"I really tried to stop watching television," the boy responded meekly.

"Well, guess what?" said Thomas. "I haven't smoked a cigar. We still have our deal." And Thomas hasn't smoked one since.

Thomas developed one-on-one mentoring relationships with some of the young people he met. Most were teenagers. All were black. And many had life experiences remarkably similar to his own. They came from broken families. They were poor. A few were both strong athletes and strong students, just like Thomas.

In 1994 Thomas struck up a relationship with Akili West, a black teenager from Washington's south side. West's parents were divorced, and he was struggling over what to do with his life. His father wanted him to go to college, but West thought about working for a while first. Thomas sat with West for more than two hours at the Supreme Court. He told the youth about his radical days at Holy Cross, a time when he considered dropping out of school and devoting his life instead to protesting racial injustice. The turning point came, Thomas said, when he thought about how hard black Americans had fought for the freedom to get an education. Was he honoring that history by protesting instead of learning? Thomas told West that he resolved that the most important thing in his life was to get an education, to become the first member of his family to earn a college degree.

"He was trying to tell me, Look, go ahead, finish your higher education," said West. "It's important. Don't be so angry about things that are happening or have happened that it causes you to shoot off your toe to spite your foot. He said that he did let anger get to him at certain points of his life and that because of it he hadn't been able to see the next big thing on the horizon."

Speaking obliquely about his confirmation, Thomas told West that he must never "return an ill with an ill."

"You have to be better than those who have done you wrong," Thomas told him.

West left his meeting with Thomas resolved to attend college. He also left with unanswered questions. Had the abuse Thomas had taken in his life generated unspoken resentments toward black Americans? How could Thomas be sure that he didn't have any deep-seated biases against members of his own race? West left before he had an opportunity to ask. That fall he headed off to Morehouse College in Atlanta, a decision he credited to Thomas.

Early in their marriage, Clarence and Virginia decided not to have children of their own. It was more Thomas's decision than his wife's.

He was already a father to Jamal. By the time he reached the judiciary in 1990, three years after their marriage, Thomas believed he simply didn't have the time to devote to being a good father. It was a painful decision, particularly for Virginia, who had no children of her own.

The summer of 1997 Thomas's family came for a weeklong visit to Washington, an annual ritual for his mother, his sister, and an ever-changing crowd of friends and relatives. Emma Mae, Thomas's sister, had four grown children of her own and was a grandmother many times over. Thomas looked forward to the visits. His mother always arrived with a cooler packed with homemade deviled crabs, his favorite. Abraham Famble, Thomas's friend from Pin Point, also always came, and Thomas never tired of arguing with Famble over who was mightier with the Weber grill.

That particular summer the Fambles, Thomas's mother, and Emma Mae arrived with some of Emma's grandchildren, Thomas's grandnieces and nephews. One of the little children was a five-year-old named Mark, a quiet, slender boy with a fair complexion. His father, Mark Martin, was Emma's son, and his mother, Susan, was white. Mark was locked away in a Georgia prison for a probation revocation relating to a conviction for selling crack. His son and two daughters lived with their mother in a public housing development in Savannah.

Clarence and Virginia Thomas were drawn to young Mark, whose biracial background mirrored their own biracial marriage. They couldn't help but notice something else: Mark's birthdate was September 22, 1991—just the day before Anita Hill had faxed her statement to the Judiciary Committee, setting in motion the events that led to the couple's darkest moments. Almost six years later Mark arrived in their home, radiating innocence.

Virginia Thomas sat out on the back deck with Mark in her lap, just holding the boy. Thomas found things for him to do around the house. He gave him a garbage bag and told him to pick up the empty soda cans. For every few cans, Thomas promised a nickel. Mark eagerly accepted the assignment, yet it created an odd impression on the young boy. His Uncle Clarence, he thought, was a garbageman.

As Mark was leaving to return home to Savannah, Thomas gave him a football. The van carrying him away wasn't even out of the driveway when Virginia Thomas turned to her husband with a sudden suggestion: Why couldn't they raise Mark? The suggestion hung there, Virginia Thomas recalled, with little discussion. Yet within days Thomas was on the telephone to Savannah, proposing to his sister and the boy's mother that he adopt Mark and give him a home in Washington. When he told his wife, she made another suggestion. Why not take in Mark and his two siblings? They made a new offer to the children's mother: If she was willing, they would adopt all three of her children.

The thought of giving up all her children was too much for Susan Martin, but Mark was getting to an age where he was more difficult to handle. He had no father in his life. Thomas stressed that he could offer Mark his own room and a safe neighborhood. They would see to it that he got the best education available. They would never attempt to supplant her role as mother, they told Susan Martin. Mark could visit when he wanted, and she was free to visit them. They suggested they all call themselves "Team Mark," parents dedicated to giving a young boy an opportunity to grow and learn.

Thomas insisted, however, that any arrangement be made legally binding. If he were to take Mark, it would be final. Thomas had tried the arrangement once before when little Clarence, Emma's oldest child, came to live with him in the early 1980s but returned home when he became homesick for his mother. Thomas contacted Joe Bergen, a lawyer who had represented his grandfather, to draft the legal documents.

In October 1997 Mark's father was released from jail and placed on probation. In what Thomas later called a rare moment of sanity, the adults signed the adoption papers transferring Mark to Thomas's custody.

Clarence and Virginia drove to Savannah before Thanksgiving to bring Mark home with them. Susan Martin could not stop crying. The Thomases reassured her that they would take care of Mark. They told her to imagine she was in a burning building and doing the only thing she could to save her son—handing him through the open window to safety.

The three of them—Clarence, Virginia, and Mark—returned to Washington as a new family. Forty-two years earlier a scared little boy with all his belongings stuffed in a shopping bag had left his mother's home to begin a new life with his grandparents. Now history was repeating itself. At forty-nine, Thomas had seen his life come full circle.

Overnight, Clarence and Virginia Thomas plunged into a new home life governed by the needs and demands of a six-year-old. Suddenly Virginia was a parent with no day-to-day experience raising a child. What did he eat? When did he sleep? How should she talk to him? The questions were endless in the first months, and she spent hours on the phone with the boy's mother.

For the second time in his life, Clarence Thomas was a father. He enrolled Mark in private school and altered his routine to fit Mark's school schedule. He continued to go to the Court early, as he always had done, while Virginia got Mark ready for school before she went to work. Thomas returned from the Court midafternoon so he could pick Mark up from school. He outfitted his home office so he could work at home, while keeping an eye on his grandnephew.

The first few weeks and months were difficult. Mark was excitable. He seemed unable to focus on anything for more than a few moments. The Thomases worried about potential learning disabilities.

Thomas immediately imposed rigid discipline in Mark's life, applying the same zeal and determination that Myers Anderson had exerted with him. He assigned extra reading and math homework. Mark responded to the new structure and began to show improvement at school. Yet as young as he was, he didn't fully comprehend the permanence of his new life. He spoke often of his mother and father in Georgia.

Riding in the car with Virginia one day, Mark talked about what he wanted to do when he got home—not home in Virginia, home in Savannah. Virginia parked the car and went around to the backseat, where she bent down and looked Mark in the eye. You are not going home, she said. This is home. His eyes filled with tears. Suddenly, she realized Mark thought he was just visiting his uncle, the garbageman who lived in the nice, big house in Virginia.

Law clerks and close friends noticed an immediate change in Thomas after Mark's arrival. He seemed years younger. He laughed more. He moved with an energy and excitement that associates hadn't seen before. Friends remarked that Thomas seemed to be living out the life of his grandfather, shaping his grandnephew's life with the vigilance of Myers Anderson.

Yet Thomas was affectionate with the boy in a way that Anderson had never been with him. He encouraged Mark to play sports—the more the better—and he never tired of playing with his grandnephew in the backyard. He taught him how to throw a football, hit a baseball, dribble a basketball. It was as though he had suddenly been given a chance to live his childhood for the first time.

Six hundred miles away in Savannah, the world was beginning to close in on Mark Martin, Mark's father. He moved in with his mother, Emma Mae, in Pin Point after he was released from prison in October 1997 and found a job building floating docks. He'd never really learned to read or write, but he was good with his hands. He liked to make things, especially out of wood. He got paid seven dollars an hour, and there was enough work to fill a twelve-hour day. A few months into the job, however, Martin took a hard fall on the dew-soaked grass, slipping a disc in his back. He lost the marina job and was unemployed for several months. By the time he was on his feet again, the dock-building job was gone. Martin took a contracting job putting up drywall. When that ended, he got work washing cars. Having dropped out of school young, he had few options.

Born in 1969, Martin was a generation behind Thomas in Pin Point. Some of his best friends were the sons of Thomas's boyhood friends. Yet the world had changed dramatically since Thomas lived in Pin Point. In the mid-1990s the community was as well known for crack cocaine as for beautiful marshes. Young people cruised the narrow neighborhood road looking for dope. Enterprising dealers drove to Miami to buy cocaine to sell back in Pin Point, Sandfly, and other black neighborhoods.

In January 1998 Mark Martin sold five grams of crack to Rufus Anderson, a friend he'd grown up with in Pin Point. Unbeknownst to him, Anderson was working with state and federal drug agents as a confidential informant. The year before, Anderson told authorities he wanted to purge Pin Point of drugs, confessing to a ten-year addiction that had wrecked his marriage and his life. Authorities wired Anderson with a hidden tape recorder and fitted his pickup truck with a concealed video camera. Anderson's buy from Martin was recorded and videotaped. Three weeks later Anderson came back to Martin for more, buying another eleven grams of crack.

Martin needed the money. He was behind on child support payments for his two little girls, who were living with their mother. Martin knew he was playing with fire because he'd pleaded guilty to selling crack twice since 1991. At a probation hearing in 1995, a Georgia superior court judge issued him a stern warning. "I want you to understand one thing," Judge Michael Karpf told Martin. "You are on probation. You're going to be walking the straight-and-narrow path, and you take one wrong step off that path and they're going to bring you back here. So, your days of running the streets are over."[1] Under drug laws, a third conviction could send Martin to prison for a very, very long time. He sold no more drugs to Rufus Anderson after February 1998. But the trap had already been sprung.

On August 19 drug agents arrested Martin and a dozen other suspects in a predawn raid. All had sold crack to Rufus Anderson, the undercover informant. All had been friends since childhood. All were from Pin Point or Sandfly. Mark Martin was charged in U.S. District Court with two counts of selling a controlled substance. Because of his prior convictions, he was held without bail.

At his trial in January, Martin took the stand in his own defense, telling the jury about his losing struggle to hold a steady job. He said he had had no intention of resuming the drug trade and had sold drugs to Rufus Anderson only because Anderson kept pestering him about it. "I gave in because I just lost my job," he told the jury. "I needed to pay some bills." Martin attempted to tell the jury about how his uncle, a Supreme Court justice, had repeatedly urged him to take responsibility for his family and put his life back together after his previous arrests. But the judge ruled the testimony out of order.

The jury convicted Martin on both counts after a one-and-a-half day trial. Because of his two prior convictions, the judge had no choice but to sentence Martin to the terms laid out in federal sentencing guidelines: thirty years in federal prison, with no chance of parole. Martin was twenty-nine years old. He'd be out just before his sixtieth birthday.

Thomas was bitterly disappointed in his nephew, friends said. On one occasion, his friend Abraham Famble recalled, Thomas said his only hope was that young Mark might get a better chance in life.

"He didn't want little Marky to be like his dad," Famble said.

Although Thomas had once fretted over managing the dual responsibilities of fatherhood and the Supreme Court, he learned he had enough time for both. He often took Mark with him to the Court, where he was free to roam around on his own. Mark's youth proved a welcome tonic in the traditionally stuffy Court atmosphere. Justice

Souter encouraged Mark to drop by when he was in the building and gave him books to read. Thomas's clerks often doubled as baby-sitters when Thomas was traveling or he and Virginia went out. The weekly luncheons of the justices invariably included a story from Thomas about Mark's latest triumphs on the basketball court or in school.

Yet as different as Mark's life was living with a Supreme Court justice, Thomas also wanted to maintain some semblance of normalcy for his adopted grandnephew. He wanted to take him camping and on summer vacations, to show him life outside Washington. He wanted the same kind of normalcy himself. The thing he hated most about the Supreme Court was his loss of anonymity and how his visibility complicated even simple things like traveling.

In November 1999 Thomas flew to Phoenix to give a speech and accept an award. With a couple of hours to kill in the afternoon, he excused himself and paid a visit to Desert West Coaches, one of the city's top dealers in luxury motor homes. He walked around the lot, examining the giant, eight-wheel behemoths. A used forty-foot Prevost, the Cadillac of motor homes, caught his eye. A thick band of steel encased the bus. Painted black-and-orange flames sizzled around the sides. Thomas kicked the tires and climbed aboard. He told Desert West owner Wayne Mullis he wanted to buy it but had to check with his wife first. "Tell you what," said Mullis. "You shake my hand, give me your word, I'll hold this bus."

Two hours later Thomas was back on the lot with Virginia, who gave the bus her blessing. After haggling over the price, Thomas was the owner of a forty-foot moving fortress. One small detail remained. Neither Clarence nor Virginia Thomas knew how to drive it, so Mullis agreed to give them lessons.

Thomas had spent months carefully researching what kind of motor home he wanted. He pored over specifications, comparing sizes and engine horsepower. New models, some costing more than $1 million, were out of his price range, so he focused on used buses, a specialty of Desert West in Phoenix. Thomas chose a 1992 model, featuring an on-board kitchen and bathroom, and bed space for up to four people.

Thomas yearned to travel more, and the bus suited his style perfectly because it was self-contained. There were no airports or hotels to deal with. He could come and go as he pleased, and not be tied to someone else's schedule. As with most aspects of his life, there were also ties to Myers Anderson, who talked about having a bus when Thomas was young. Because of the hostility blacks faced when traveling, Anderson rarely took long road trips. With a motor home, Anderson said, the family could travel and see the country.

In mid-December 1999 the Thomases bought a one-way plane ticket to Phoenix. Mullis spent hours with them on his bus lot, showing them how to back up, park, change gears, and handle their new set of wheels.

On the way home Thomas called Mullis and put his wife on the line. "I'm going down U.S. 80 and I'm going seventy miles an hour and it's a snowstorm and I'm driving this bus!" she gleefully shouted into the receiver. "What do you think about that, Mr. Bus Driver Instructor?"

Several weeks after Thomas returned to Washington from his first cross-country bus trip, his brother, Myers, was home visiting their mother in Savannah for a weekend in mid-January. He ate her deviled crabs and visited old friends in Savannah. An accountant by training, Myers Thomas was now president of a small Louisiana company. He lived a prosperous life in New Orleans with his wife, a Georgia native, and their two children. As he left Savannah, Thomas told his mother he planned to drive to Liberty County to visit his grandparents' grave sites before driving across Georgia and Alabama to New Orleans.

The following Sunday before church, Myers and his wife went out for a morning jog, leaving their adolescent children at home. Their father never came back. In the middle of the jog, he collapsed with a massive heart attack and died that morning. He was fifty years old.

For the third time, Thomas returned to Georgia to make funeral arrangements. Myers was to be buried next to their grandparents in Liberty County, the very ground where he'd been standing just a week before he died. The funeral was held at Saint Benedict's Church, across the street from where Myers and Clarence Thomas had attended grade school.

At the wake Thomas spent long minutes gazing down at his brother's lifeless body. Being on the Supreme Court was meaningless, he thought to himself. What mattered was not what you left behind but how you lived your life in the moment. Faith, family, and friends—those were the important things in life, Thomas concluded. "Not who is the smartest, not the score we have at the end of the day, not who has written the most opinions or the most dissents, not who is the most akin to Oliver Wendell Holmes," he said later. "But whether or not we are akin to someone like my brother in spirit. Have we done the principled thing? Have we done the right things in our faith and by our families and by our friends?"[2]

His brother's death plunged Thomas into a long period of mourning. Myers was Thomas's link to his grandparents and to their childhood. Although separated by more than a year, Thomas thought of his brother as his twin. For sixteen years they were almost inseparable. They shared the unspoken bond of family and common experience. Myers Thomas had known Clarence Thomas probably better than anyone else. Now he was gone.

"I think that if he had the power to go into the past and change one personal event, the event would be the death of his brother," says G. Michael Fenner, a friend. "That hit him hard—harder perhaps than the confirmation hearings."

Thomas resolved to become even more involved in the lives of his brother's two children, and he took a special interest in ten-year-old

Derek, who was deeply affected by his father's death. Myers's death also put Thomas in much closer contact with his mother. While he was alive, Myers had been responsible for looking after their mother's financial needs and their grandfather's house. With his death, that responsibility fell to Thomas.

Thomas also continued to throw himself into new mentoring relationships with young people, often because he simply didn't know how to say no. In May 2000 teachers from a school in Staunton, Virginia, approached Thomas after an hours-long session with the school's fifth-grade class. Guidance counselor Pat Lynn pleaded with Thomas to help one of her top students, a young black boy. Mario Scott was a brilliant student and a gifted athlete. He lived in Staunton with his father, a short-order cook, and his father's girlfriend. Lynn told Thomas she feared Mario might drop out of school and she was trying to get him enrolled in a free private school in Pennsylvania for gifted students. Could Thomas pull some strings? Thomas said he had no contacts at the Pennsylvania school but knew people at the military academy in Virginia that Jamal had attended.

Weeks later Lynn received a call from a Washington attorney who said Mario had been admitted to the military school in the fall. An anonymous donor would pay the school's nineteen-thousand-dollar-a-year tuition and board for Mario through the twelfth grade. Thomas had found the donor and arranged the financing.

Thomas has helped other Marios over the years but actively discourages publicity about the assistance. In public, he's spoken about how few black role models were available to him as a child growing up in Savannah.

In several speeches, he has cited a passage from Charles Dickens's *Bleak House* that appears to capture his approach to helping people. In the novel, a character named Mrs. Jellyby operates a charity for far-flung causes in Africa, while a bevy of dirty, neglected children runs wildly about her house. For Thomas, charity begins at home, one child at a time. He wants to help people he can see and touch, and whose progress he can follow with the same vigilance that his grandfather exerted with him.

Mario Scott enrolled at Fork Union Military Academy in Virginia in the fall of 2000. Thomas wrote Mario letters of encouragement, and invited him to a lunch at James Madison University six months later. At the lunch, he put his arm around the slender young boy and playfully mauled the top of Mario's head with his strong hand. "You remind me of my little guy," said Thomas.

In May Thomas flew to his native Savannah to headline a luncheon for the local bar association. The timing of the event, two months before the end of the term, was terrible. But the luncheon was to honor Joe Bergen, who had represented Anderson at a time when few white lawyers would take black clients. Bergen was also the attorney who had arranged Mark's adoption in 1997.

Toward the end of his speech, Thomas began to thank Bergen for helping with Mark's adoption. Partway through his comments, Thomas locked eyes with his own mother and stopped in midsentence. An involuntary shudder shook his entire body. He closed his eyes and pinched the bridge of his nose, trying to hold back the tears and intense emotion boiling up inside. Nearly half a minute passed before he spoke again. "Boy, that's embarrassing," he said.

In the uneasy silence, someone called out softly, "You're home."

"I apologize for the interruption," Thomas said, still unable to continue and dabbing at his eyes. Finally, he gave up. "Suffice it to say that I thank God for Joe Bergen."

It was a rare moment of exposure for such a private figure. In an instant, the uncompromising tough guy dissolved into an emotional, sensitive man. Talking about rescuing his grandnephew touched the anguish Thomas still feels over losing his own father and a portion of his childhood. The painful lessons of loss and abandonment came early for Thomas, and they repeated themselves with the deaths of his grandparents and his brother.

The moment of melancholy lasted only a few moments. After the speech, Thomas angrily rebuffed a news reporter's request for an interview. "Why would I do that?" he snapped. The news media had a chance to tell his story, he said, but "blew it." Thomas instructed Bergen to disclose no details about Mark's adoption, telling his family lawyer that it was "none of the press's damn business."

Shortly after his visit to Savannah, Thomas became embroiled in a controversy that struck very close to home. Harlan Crow, a wealthy Texas businessman, approached a local foundation raising money to renovate Carnegie Library, the formerly segregated library where Thomas had learned to read as a boy. Crow offered $150,000 to complete the renovations on one condition: He wanted Carnegie renamed for his friend Clarence Thomas.

Several black members of the Savannah-Chatham County library board, a separate body, were outraged when they learned of the condition. "Clarence Thomas has never cared anything about black folks," declared board member Robert Brooks. "He looks upon his race as Hitler looked upon Jews, and I call him Judas because he sold his people out."[3] Brooks organized several community groups to campaign against the terms of the donation.

The library foundation presented Crow with a different proposal. His donation would go toward construction of a new wing at Carnegie named after Thomas, but the library's name would not change. Crow accepted.

In September Thomas returned from a weekend reunion in Missouri with Conception classmates, refreshed and ready for the start of the new Supreme Court term. Two days later terrorists struck a historic blow against the United States, killing thousands in attacks on the World Trade Center and the Pentagon. For Thomas, the attacks hit very close to home. Flight 77, which slammed into the Pentagon, carried Barbara Olson, a close friend of the Thomases'. Olson, a conservative author, was married to United States solicitor general Ted Olson. She had worked hard supporting Thomas's nomination to the Supreme Court, and her death personalized the September 11 tragedy for both Thomas and his wife. Thomas delivered Olson's eulogy four days after the tragedy.

Thomas finished his eleventh year on the Court at the end of June the following year. One of the term's biggest cases was his. Writing for the Court majority, Thomas upheld the constitutionality of a mandatory drug-testing policy in an Oklahoma school district. The decision made no pronouncements on the wisdom of the school's policy but declared it was a reasonable measure to deter drug use among students.

Thomas also weighed in on a major decision that upheld the constitutionality of an Ohio program that used public money for religious-school tuition. The case was a landmark decision for proponents of school vouchers. For Thomas, the case touched three topics close to his heart: the Catholic church, minority children, and education. Catholic parochial schools received the bulk of the tuition voucher money under the Ohio program, and most of the students at the schools were black or minority. Test scores showed that the children at the schools were performing better than those at the public schools.

"Just as blacks supported public education during Reconstruction, many blacks and other minorities now support school choice programs because they provide the greatest educational opportunities for their children in struggling communities," Thomas wrote.

"Staying in school and earning a degree generates real and tangible financial benefits, whereas failure to obtain even a high school degree essentially relegates students to a life of poverty and, all too often, of crime," he continued. "The failure to provide education to poor urban children perpetuates a vicious cycle of poverty, dependence, criminality, and alienation that continues for the remainder of their lives. If society cannot end racial discrimination, at least it can arm minorities with the education to defend themselves from some of discrimination's effects."[4]

In June 2002 Thomas traveled south to attend the annual conference of the Georgia Bar Association on Amelia Island, Florida, just across the Georgia border. Thomas flew first to Tampa, near a maintenance depot where he kept his beloved bus over the winter for annual repairs. John Yoo, a former clerk, came along to keep him company and help him drive the bus back to Washington. They got a late start and didn't reach Amelia Island until evening. Thomas checked in to a suite on the eighth floor of the Amelia Island Plantation, a pricey resort built along miles of picturesque Florida beach. The June weather was hot and beginning to become humid, but the evenings were still cool and refreshing.

Thomas's room opened to a broad balcony overlooking the beach and the ocean. To the north lay the Georgia coastline, dotted with barrier islands that led all the way to Savannah. Somewhere in the distance was Liberty County. Two hundred years ago the islands and adjoining lowlands produced the finest sea-island cotton in the world, grown by tens of thousands of Georgia slaves. Amelia Island was also once a former slave plantation. The notion struck Thomas as funny: he, a black man, staying on the plantation.

From his balcony, Thomas could also look north toward American Beach. As a boy, Thomas swam at the segregated beach—the only one available to blacks—after a long, hot bus ride from Savannah. How much had changed in the country and in his own life since then. Who could have imagined he would be where he was some forty years later?

Before his speech to the conference the next day, he scooted off to visit American Beach and have lunch in the old city of Fernandina. At the reception in his honor that night, Thomas posed for photographs and shook hands with a steady stream of well-wishers. Just two weeks shy of his fifty-fourth birthday, traces of the short, wiry athlete who threw a football seventy-five yards were hard to see. The once-narrow waistline was broad and round, the face full and fleshy. Only the shoulders and biceps, sculpted by regular weight lifting, suggested the athleticism of Thomas's youth. The curly black hair was almost completely white. The laugh remained.

Thomas took the podium that evening looking refreshed and relaxed. He reminisced about driving the old Ford tractor for his grandfather on the farm in Liberty, cursing his grandfather's relentless work demands, and dreaming of a life far removed from manual labor. He spoke of his youthful longing to return to Savannah to practice law, a dream that ended when he joined the federal judiciary in 1990. He paused to say a few words about Bobby Hill, the Savannah lawyer who offered him a job in 1973. Hill had died in 2000 after a life of accomplishments and failures. He battled drug addiction and scandal, disbarred in 1984 after mishandling client funds. "I'm convinced if we were to go back and look at the numbers of people who were positively influenced by him, we would be quite amazed," Thomas said of his former idol. "What he dreamed of doing supplied young people like me, especially me, with a hope in my own life, and a hope that one day I could be a decent lawyer and could have some impact on our society."

Thomas talked about how happy he was on the Court, a place where people managed to disagree without being disagreeable. "I have had my moments when I was upset because I couldn't get a majority," he said. "I have been frustrated from time to time. I have gone back to my chambers and just thrown up my hands—'What are these people thinking?' But I will still say, in all that frustration, that we are the envy of the world, and for good reason. We don't have revolutions when we change our governments. We don't have wars when we have differences of opinions. We have a system to solve our disputes, to re-solve our disputes. And it works.

"I would love for you to be a fly on the wall at our Court. I wish you could see it. Perhaps we would be a little embarrassed. Perhaps we wish we could remain in secret, but the bottom line, and I truly be-lieve this, after you see how the sausage is made, you will come away with more pride, and more pride about our country and our judicial system."

Over the years, Thomas said he had become more idealistic about America, not less.

"The longer I've had the obligation to interpret our Constitution and our laws, the more I believe in our system and in our government, the more I believe in our Constitution, the more I believe in our framers, the more faith I have in the amendments, and the more faith I have in our citizens. So, I'm probably reaching the point where I'm an apologist for the Court, for the Constitution, and for our country. And that's an odd thing for a young kid who grew up, up the road here, on Route 17, wondering where all those cars were going when they were headed south. . . .

"So I say that as someone who has suffered the exclusion, and someone who would have all sorts of bases to be angry, and all sorts of justifications. This system is the finest available in the world, and I'm more idealistic. And it is this system that will solve the problems, not the ones we see being generated around the world. So, yes, with that background I am an apologist for this system. Because I believe that had those judges in our country been true to it, the problems that I lived under would never have existed. And the way to solve it is not to throw out this system that would have worked then, but to make it work as it should have."

EPILOGUE

Several months after his speech in Florida, Thomas wrote to express his frustration that yet another book about his life would soon be forthcoming. "I do not understand this interest in me," he said. "At bottom, I merely tried to do as my grandfather advised: make the best of the hand that had been dealt me. Perhaps some are confused because they have stereotypes of how blacks should be and I respectfully decline, as I did in my youth, to sacrifice who I am for who they think I should be."

The handwritten note included a short postscript on the other side of the stiff, cream-colored card: "I am writing my own memoirs so that I might give some clarity to my life," he said. "It is difficult for many reasons, but it must be done, especially given that so much has been written about me that has no basis in reality."

Six months later Thomas landed a $1.5 million advance for the book. The size of the deal surprised him, but he couldn't help but enjoy the irony. Without Hill's allegations a decade earlier and the trashing from liberals ever since, he never would have commanded such an extraordinary sum for his life story. Thomas and Virginia made big plans to spruce up their house and buy all the furnishings they hadn't been able to afford when they moved in.

Not a bad tonic for all the heartache and bitterness. Thomas would have the last laugh, the last word—and plenty of money.

As Thomas himself might say, "Not bad, for a starvin' man."

NOTES

CHAPTER 2: SANDY WILSON

1. Information about Sandy Wilson's early life was culled from the estate records of Josiah Wilson, the property and tax records of Liberty County, and census records from 1800 to 1880 on file at the Georgia Archives, Morrow, Georgia.

2. Bulloch died in 1849 near Atlanta, acquiring a footnote in history years later when his daughter gave birth to Theodore Roosevelt, the nation's twenty-sixth president.

3. See Philip D. Morgan's scholarly account of property ownership in coastal Georgia and South Carolina in "The Ownership of Property by Slaves in the Mid-Nineteenth-Century Low Country," *Journal of Southern History* 49, no. 3 (August 1983).

4. Liberty County, Georgia, Case Files, Southern Claims Commission. On file at the Georgia Archives.

5. Ibid.

6. Ibid.

7. Ibid.

8. Remnants of the task system survived in the work habits of Myers Anderson. Anderson always wanted to be paid by the job, so he could control when and how it was finished. He thought hourly wages were a prescription for laziness and remained opposed to a minimum-wage law until his death, according to Thomas.

9. James Stacy, *History and Published Records of the Midway Congregational Church Liberty County, Georgia* (Spartanburg, S.C.: Reprint Company, 2002), p. 171.

10. Liberty County Case Files, claim of Samuel Osgood.

11. Ibid.

12. Ibid.

13. Ibid.

14. Ibid.

15. Ibid.

16. Obituary of O. W. Hart, 1874, courtesy of Carol Van Cleef.

17. In her journal of Union occupation of Liberty County, Mary S. Mallard says thousands of freedmen and -women lived in Bryan County in the chaotic two years following the war. Robert M. Myers, ed., *The Children of Pride* (Binghamton, N.Y.: Vail-Ballou Press, 1972).

18. Ira Berlin et al., eds., *Free at Last: A Documentary History of Slavery, Freedom and the Civil War* (Edison, N.J.: Blue & Grey Press), p. 314.

19. Politically, the years following Reconstruction were also those when white, southern Democrats codified Jim Crow segregation and stripped free blacks of their constitutional right to vote. In Liberty County the effort to disfranchise the newly freed slaves began almost immediately after the war. White Democrats in the Georgia General Assembly, claiming the state constitution did not specifically allow black lawmakers, refused to seat Golden, Campbell, and the 28 other black legislators elected in 1868. They were admitted only after Congress expelled Georgia from the Union for the second time. In McIntosh County, Campbell was ultimately sentenced to a work camp on trumped-up charges and a sham trial that federal authorities in Atlanta ultimately determined was a racially and politically motivated "disgrace." See Russell Duncan, *Freedom's Shore: Tunis Campbell and the Georgia Freedmen* (Athens: University of Georgia Press, 1986), p. 106.

20. Liberty County Case Files.

CHAPTER 3: MYERS ANDERSON

1. Laticia Allen's name is spelled differently on different records. The spelling used here came from Rosa Allen, her niece.

2. Justice Thomas says Anderson never knew his real birth year. Both his death certificate and headstone list it as 1907. Social Security Administration records indicate it was 1910. The 1920 census recorded his age as nine, suggesting the later birth year.

3. Writers' Program, *Drums and Shadows: Survival Studies Among the Georgia Coastal Negroes* (Athens: University of Georgia Press, 1940), p. 82.

4. Chatham County Superior Court files.

CHAPTER 4: BLACK AND WHITE

1. Speech to University of Louisville, Louisville, Ky., 11 September 2000.

2. *Savannah Herald*, 11 March 1961.

3. *The Crisis*, March 1983.

4. University of Louisville speech.

CHAPTER 5: THE ISLE OF NO HOPE

1. *Savannah Morning News*, 25 January 1964.

2. *Atlanta Constitution*, 24 May 1963.

3. *Esquire*, July 1998.

4. Report by the California Governor's Commission on the Los Angeles Riots, *Violence in the City—An End or a Beginning?* 2 December 1965.

5. *The Pioneer*, Fall 1966.

6. *Washington Post*, 17 July 1983.

CHAPTER 6: "READY FOR WAR"

1. *The Crusader*, 26 April 1986, College of the Holy Cross Archives.

2. Speech to College of the Holy Cross, Worcester, Mass., 3 February 1994.

3. Anthony J. Kuzniewski, *Thy Honored Name: A History of the College of the Holy Cross, 1843–1994* (Washington, D.C.: Catholic University of America Press, 1999), pp. 370–99.

4. Files of the Black Student Union, College of the Holy Cross Archives.

5. Stokely Carmichael and Charles V. Hamilton, *Black Power: The Politics of Liberation in America* (New York: Random House, 1967), p. 44.

6. *Worcester Telegram*, 18 April 1969.

7. Ralph Ellison, *Invisible Man* (New York: Random House, 1994), p. 3.

8. Report of the President's Commission on Campus Unrest (Washington: U.S. Government Printing Office, 1970), p. 39.

9. Files of the Black Student Union, College of the Holy Cross Archives.

10. *The Crusader*, 19 December 1969.

11. *Ibid.*, 23 April 1971.

12. Speech to National Bar Association, Memphis, Tenn., 29 July 1998.

13. Claude McKay, "If We Must Die," Archives of Claude McKay, Carl Cowl, administrator.

14. Paul Laurence Dunbar, *The Collected Poetry of Paul Laurence Dunbar*, ed. Joanne M. Braxton (Charlottesville: University Press of Virginia, 1993).

15. *Crossroads*, May 1971.

CHAPTER 7: IVY LEAGUE FAILURE

1. Judge Louis Pollak, Oral Tapes and Transcripts of the Griswold-Brewster History Project, Yale University, 1990–1993. Record Unit 217, Manuscripts and Archives, Yale University Library, p. 23.

2. Kingman Brewster, Jr., Griswold-Brewster History Project. p. 21.

3. Speech to George Mason University Law School, Fairfax, Va., 29 May 1999.

CHAPTER 8: "NOT BAD FOR A STARVIN' MAN"

1. *Wall Street Journal*, 13 September 1989.

2. Thomas Sowell, *Race and Economics* (New York: David McKay, 1975), p. 238.

CHAPTER 9: BLACK CONSERVATIVE

1. *Washington Post*, 16 December 1980.

2. Anita Hill, *Speaking Truth to Power* (New York: Doubleday, 1997), p. 59.

3. *Ibid.*, pp. 57–58.

4. *Ibid.*, pp. 59–60.

CHAPTER 10: THE DUNGEON

1. EEOC videotape, EEOC Library, Washington, D.C., 1991.

2. Senate Committee on Labor and Human Resources, *Oversight of the Equal Employment Opportunity Commission: Hearing before the Committee on Labor and Human Resources*, 97th Congress, 2nd sess., 15 June 1982.

3. Speech to Heritage Foundation, Washington, D.C., 17 June 1987.

4. Laurence I. Barrett, *Gambling with History: Ronald Reagan in the White House* (New York: Doubleday, 1983), p. 416.

5. Michael K. Deaver, *A Different Drummer: My Thirty Years with Ronald Reagan* (New York: HarperCollins, 2001), p. 97.

6. EEOC videotape.

7. House Committee on Education and Labor, *Oversight Hearing on the EEOC's Enforcement Policies: Hearing before the Subcommittee on Employment Opportunities of the Committee on Education and Labor*, 98th Congress, 2nd sess., 14 December 1984.

8. Speech to high school students, Supreme Court, 13 December 2000.

9. *Esquire*, February 1987.

10. Speech to International Association of Human Rights Agencies, Philadelphia, 11 July 1983.

11. Speech to Missouri Human Rights Conference, Columbia, Mo. 20 May 1983.

12. Speech to the Bureau of the Census, Equal Employment Opportunity Awareness Week, Washington, D.C., 2 May 1983.

13. Speech to University of Virginia, Charlottesville, 19 October 1983.

14. *The Crisis*, March 1983.

15. *Newsweek*, 31 October 1983.

16. *Washington Post,* 25 October 1984.

17. *Ibid.,* 12 March 1984.

18. *The Fountainhead,* MGM/UA Home Video, 1949.

CHAPTER 11: FAST TRACK

1. Sheldon Goldman, *Picking Federal Judges* (New Haven: Yale University Press, 1997), p. 297.

2. Speech to American Bar Association, Washington, D.C., 9 July 1985.

3. House Committee on Education and Labor, *Oversight Hearing on EEOC's Proposed Modifications of Enforcement Regulations, Including Uniform Guidelines on Employee Selection Procedure: Hearing before the Subcommittee on Employment Opportunities,* 99th Congress, 1st sess., 2 October 1985.

4. Speech to Compton Community College, Compton, Calif., 14 February 1986.

5. Speech to Georgetown Law Center EEO Symposium, Washington, D.C., 20 February 1986.

6. *Reason,* November 1987.

7. *Plessy v. Ferguson,* 163 U.S. 537 (1896).

8. Speech to Georgia Southern College, Statesboro, Ga., 24 February 1987.

9. Speech to Cato Institute, Washington, D.C., 23 April 1987.

CHAPTER 12: VIRGINIA

1. Speech to Heritage Foundation, Washington, D.C., 17 June 1987.

2. *Wall Street Journal,* 8 July 1987.

3. Speech to Pacific Research Institute, San Francisco, 10 August 1987.

4. *Los Angeles Times,* 25 February 1988.

5. Speech to Palm Beach Chamber of Commerce, Palm Beach, Fla., 18 May 1988.

CHAPTER 13: RELUCTANT JUSTICE

1. Senate Committee on the Judiciary, *The Complete Transcripts of the Clarence Thomas–Anita Hill Hearings,* ed. Anita Miller (Chicago: Academy of Chicago, 1994), pp. 13–17.

2. Letter to Senator Joe Biden, 31 October 1989, files of the Alliance for Justice.

3. Juan Williams, *Thurgood Marshall: American Revolutionary* (New York: Random House, 1998), p. 390.

4. Tom Jipping, "'Judge Thomas Is the First Choice': The Case for Clarence Thomas," *Regent University Law Review* 12, no. 2, p. 400.

5. Associated Press, 28 June 1991.

6. Speech to Eighth Circuit Judicial Conference, Kansas City, Mo., 17 July 1999.

7. Eighth Circuit Judicial Conference speech; Jim McGrath, ed., *Heartbeat: George Bush in His Own Words* (New York: Scribner, 2001), p. 138.

8. Timothy M. Phelps and Helen Winternitz, *Capitol Games: Clarence Thomas, Anita Hill and the Story of a Supreme Court Nomination* (New York: Hyperion, 1992), p. 18.

9. *Washington Times*, 3 July 1991.

10. *St. Petersburg Times*, 6 July 1991; Associated Press, 5 July 1991.

11. *NBA Magazine*, October 1991.

12. *New York Times*, 13 July 1991.

13. Notes of the director, files of the Alliance for Justice.

14. NAACP, *A Report on the Nomination of Judge Clarence Thomas as Association Justice of the United States Supreme Court*, p. 10.

15. John C. Danforth, *Resurrection: The Confirmation of Clarence Thomas* (New York: Viking, 1994), p. 24.

16. Senate Committee on the Judiciary, *Nomination of Judge Clarence Thomas to Be Associate Justice of the Supreme Court of the United States: Hearings Before the Committee on the Judiciary*, 102nd Congress, 1st sess., 1991 (Washington: U.S. Government Printing Office), pt. I, pp. 527–30.

17. *Wall Street Journal*, 28 July 1991.

18. "Lessons Learned from the Nomination," files of the Alliance for Justice.

19. *New York Times*, 10 September 1991.

CHAPTER 14: TRIAL OF HIS LIFE

1. Senate Committee on the Judiciary, *Nomination of Judge Clarence Thomas to Be Associate Justice of the Supreme Court of the United States: Hearings Before the Committee on the Judiciary*, 102nd Congress, 1st sess., pt. I, p. 369.

2. *Ibid.* p. 142.

3. *Ibid.* p. 235.

4. Arlen Specter, *Passion for Truth: From Finding JFK's Single Bullet to Questioning Anita Hill to Impeaching Clinton* (New York: William Morrow, 2000), p. 347.

5. Danforth, *Resurrection*, p. 29.

6. Senate Judiciary Committee, *Nomination*, pt. 3, p. 141.

7. *Report of Temporary Special Independent Counsel Pursuant to Senate Resolution 202* (Washington: U.S. Government Printing Office, 1992), Document 102–20, pt. I, p. 16.

8. *Report of the Temporary Special Independent Counsel*, pt. 2, pp. 30–31.

9. Danforth, *Resurrection*, p. 31.

10. *Ibid.*, p. 34.

11. *Ibid.*, pp. 35–36.

12. *Ibid.*, p. 41.

13. Specter, *Passion for Truth*, p. 347.

14. *Ibid.*, p. 346.

15. Anita Hill, *Speaking Truth to Power* (New York: Doubleday, 1997), p. 120.

16. *Report of Temporary Special Independent Counsel*, pt. 1, p. 22.

17. *Ibid.*, p. 20.

18. Specter, *Passion for Truth*, p. 349.

19. Danforth, *Resurrection*, pp. 48, 213–214.

20. Senate Committee on the Judiciary, *Transcript of the Executive Session of the Senate Committee on the Judiciary*, 102nd Congress, 1st sess., 27 September 1991.

21. *Report of Temporary Special Independent Counsel*, pt. 1, pp. 22, 26.

22. *Ibid.*, p. 30.

23. Danforth, *Resurrection*, p. 54.

24. *Ibid.*, p. 53.

25. *Report of the Temporary Special Independent Counsel*, pt. 1, p. 38.

26. Danforth, *Resurrection*, p. 57.

27. *People*, 11 November 1991.

28. *Congressional Record*, 102nd Congress, 1st sess., 8 October 1991, pt. 25891.

29. Jim McGrath, ed., *Heartbeat: George Bush in His Own Words* (New York: Scribner, 2001), p. 298.

30. Danforth, *Resurrection*, p. 108.

31. *Ibid.*, p. 124.

32. Senate Committee on the Judiciary, *The Complete Transcripts of the Clarence Thomas-Anita Hill Hearings*, ed. Anita Miller (Chicago: Academy of Chicago, 1994), pp. 130–134.

33. *Ibid.*, p. 26.

34. Barbara Bush, *Barbara Bush: A Memoir* (New York: Scribner, 1994), p. 436.

35. Specter, *Passion for Truth*, p. 372.

36. *Complete Transcripts*, p. 96.

37. Specter, *Passion for Truth*, p. 362.

38. *Complete Transcripts*, pp. 117–118.

39. *Ibid.*, p. 394.

40. *Ibid.*, p. 294.

41. *Ibid.*, p. 472.

42. *Senate Journal*, 102nd Congress, 1st sess., 15 October 1991.

CHAPTER 15: WOUNDED BEAR

1. Juan Williams, *Thurgood Marshall: American Revolutionary*, (New York: Random House, 1998), p. 394.

2. *Reason*, February 1992.

3. Orrin Hatch, *Square Peg: Confessions of a Citizen Senator* (New York: Basic Books, 2002), p. 160.

4. *Hudson* v. *McMillian*, 503 U.S. 1 (1992).

5. *Georgia* v. *McCollum*, 505 U.S. 42 (1992).

6. *United States* v. *Fordice*, 505 U.S. 717 (1992).

7. *St. Petersburg Times*, 30 May 1993.

8. Henry J. Abraham, *Justices, Presidents and Senators* (New York: Rowman & Littlefield, 1999), p. 317.

9. *Emerge*, 30 November 1993.

10. 512 U.S. 874 (1994), *Holder* v. *Hall*.

11. *Ibid.*

CHAPTER 16: REHABILITATION

1. *Washington Afro-American*, 29 October 1994.

2. CNBC, 12 October 2002.

3. Speech to Eagle Forum, 9 November 1996.

4. *United States* v. *Lopez*, 514 U.S. 549 (1995).

5. *Washington Post*, 29 May 1995.

6. *Capitol Square Review Bd.* v. *Pinette*, 515 U.S. 753 (1995).

7. *Adarand Constructors* v. *Pena*, 515 U.S. 200 (1995).

8. *Missouri* v. *Jenkins*, 515 U.S. 70 (1995).

9. *Rosenberger* v. *Rector and Visitors of the University of Virginia*, 515 U.S. 819 (1995).

10. *Tower Topics*, Winter 2001.

11. *Washington Post*, 9 June 1996.

12. *Emerge*, 30 November 1996.

13. *Ibid.*, March 1996.

14. John F. Callahan, ed., *The Collected Essays of Ralph Ellison* (New York: Random House, 1995), p. 163.

15. Speech to National Bar Association, Memphis, 29 July 1998.

16. Speech to University of Montana, Missoula, 13 April 1999.

17. Speech to Ashland University, Ashland, Oh., 5 February 1999.

18. Speech to Goldwater Institute, Phoenix, Ariz., 19 November 1999.

19. Speech to high school students, Supreme Court, 13 December 2000.

20. Speech to University of Louisville, Louisville, Ky., 11 September 2000.

21. Abraham, *Justices, Presidents, and Senators*, p. 313.

22. Douglas Scott Gerber, *First Principles: The Jurisprudence of Clarence Thomas* (New York: New York University Press, 1999), p. 137.

23. Speech to American Enterprise Institute, Washington, D.C., 13 February 2001.

CHAPTER 17: CHILDREN

1. Chatham County Superior Court files, 21 February 1995.

2. Speech to Federal Bar Association, Tampa, Fla., 4 April 2000.

3. *Savannah Morning News*, 14 July 2001.

4. *Zelman v. Simmons-Harris*, 536 U.S. 639 (2002).

INDEX